ETHICAL IMPERIALISM

ETHICAL IMPERIALISM

Institutional Review Boards and the Social Sciences, 1965–2009

ZACHARY M. SCHRAG

The Johns Hopkins University Press
Baltimore

Printed in the United States of America on acid-free paper
2 4 6 8 9 7 5 3 1

The Johns Hopkins University Press
2715 North Charles Street
Baltimore, Maryland 21218-4363
www.press.jhu.edu

Library of Congress Cataloging-in-Publication Data
Schrag, Zachary M.
Ethical imperialism : institutional review boards and the social sciences, 1965–2009 / Zachary M. Schrag.
p. cm.
Includes bibliographical references and index.
ISBN-13: 978-0-8018-9490-9 (hardcover : alk. paper)
ISBN-10: 0-8018-9490-5 (hardcover : alk. paper)
1. Social sciences—Research—Moral and ethical aspects—United States. 2. Institutional review boards (Medicine)—United States—History. 3. Privacy—United States—History. I. Title.
H62.5.U5S37 2010
300.72—dc22 2009043789

A catalog record for this book is available from the British Library.

Special discounts are available for bulk purchases of this book. For more information, please contact Special Sales at 410-516-6936 or specialsales@press.jhu.edu.

The Johns Hopkins University Press uses environmentally friendly book materials, including recycled text paper that is composed of at least 30 percent post-consumer waste, whenever possible. All of our book papers are acid-free, and our jackets and covers are printed on paper with recycled content.

For my parents

CONTENTS

PREFACE

Do I need official permission to interview people about their lives and work? That was the question facing me in the summer of 2000, when I received a grant from the National Science Foundation to support my research on the history of the Washington Metro. If I involved "human subjects" in my research, the grant required me to get approval from something called an institutional review board, or IRB. Since the definition of human subjects research seemed to include the oral history interviews I was conducting, I accordingly gained approval from the IRB at Columbia University, where I was a student. I kept that approval active until I had completed my work, published in 2006 as *The Great Society Subway: A History of the Washington Metro*. Meanwhile, I took a job at George Mason University, where the orientation for new faculty included a stern warning about the need to get approval for human subjects research. Somewhat reluctantly, I sought and received the approval of Mason's Human Subjects Research Board for a series of interviews on the history of riot control. At neither university was the approval process particularly difficult. I completed the mandatory online training, filled out some forms, and, after a certain interval, received permission to go ahead. At most I spent a few hours completing the required paperwork, and my research was delayed by a few weeks.

But if the injury was slight, the insult was great. Columbia demanded annual reports tabulating everyone I had spoken with—by sex and race—thus reducing distinguished architects, engineers, and statesmen to numbers on a spreadsheet and requiring me to guess the racial identities of people whom I had only spoken with on the telephone. Mason insisted that I submit sample interview questions in advance, despite my protests that since I tailor my questions to each person interviewed, sample questions would be meaningless. And both universities' training and paperwork demanded that I embrace a set of methods and ethics designed by and for medical researchers. Because they require researchers to do no harm to the subjects of their research, these ethical codes would, if taken seriously, prevent me from holding people to account for their words and deeds, one of the historian's highest duties. Had I been asked to pledge allegiance to a foreign monarch or affirm that the sun revolves around the earth, complying would have taken even less of my time, yet few would question my outrage.

Outrage alone, however, could not have sustained me through this project; this book is built on curiosity as well. On speaking with administrators and board members at Mason, I learned that they believed their decisions to be directed and constrained by a web of guidance from government agencies, along with nongovernmental Internet discussion boards, conferences, and paid consultants. I was seeing one strand of a national network of people who are convinced of the need to restrict research in the humanities and the social sciences. I found this regime so contrary to the norms of freedom and scholarship that lie at the heart of American universities that I could not imagine how it emerged. Historians love such puzzles, and when I began researching the history of IRB regulation, I did so out of love as well as anger.

I cannot disguise my frustration with the present system of regulation (for a continuing exposition of my views, see my Institutional Review Blog, www.institutionalreviewblog.com). But I approached this project with as open a mind as I could, abandoning some of my preconceptions as my research progressed. In particular, I began my investigations expecting to find that the officials who wrote today's human subjects regulations had wholly failed to consider the impact of their work on nonbiomedical research, and I was repeatedly surprised to learn how frequently (if superficially) they debated the subject. I hope all those interested in the topic, regardless of their positions, will likewise let themselves be surprised by the facts in this account.

I also write in the belief that knowledge is power. In my case, this meant the power to avoid additional IRB oversight while adhering to the policies of George Mason University. Mason defines human subjects research using the formula provided by the federal government, which in turn characterizes research as "a systematic investigation, including research development, testing and evaluation, designed to develop or contribute to generalizable knowledge." At the time I submitted my Metro and riot-control projects for IRB review, I thought that this definition covered oral history. But in January 2007, I learned that the federal Office for Human Research Protections (OHRP) itself had conducted oral history interviews. I asked if the project had received IRB approval, and an official replied that the office had not sought such approval on the grounds that "the activity was not a systematic investigation, nor was it intended to contribute to generalizable knowledge. This oral history activity was designed merely to preserve a set of individuals' recollections; therefore, this activity was not subject to IRB review." Since I saw no distinction between the interviews conducted by the OHRP and those I planned for this book, I concluded that my interviews would also be designed merely to preserve a set of individuals' recollections and were therefore not subject to IRB review. I should add that the OHRP and I interviewed some of the people who developed the current system of regulation, and none of them objected to the fact that we had not sought IRB approval for the interviews.

The gap between the actions of OHRP officials and the deeds done in their name by university administrators is emblematic of the tangled mess that is the regulation of interview, survey, and observational research. While I have my own ideas about how we can get out of that mess, I hope that everyone will benefit from a clear account of how we got into it.

My first attempt at such an account was an eleven-page memo about IRB oversight of oral history, written in March 2005. The expansion of that memo into this book was the result of constant encouragement and guidance from scholars and friends in a wide range of disciplines and institutions. The encouragement began at home, in the Department of History and Art History at George Mason University. Jack Censer—first as my chair, then as my dean—told me it was a fine thing for a newly hired assistant professor to question university policy. Roy Rosenzweig, a paragon of historians, convinced me that the questions I was asking mattered to the profession as a whole. Not long before his death, he offered wise comments on early drafts of this work, and I regret beyond words that I cannot put the finished product in his hands. Other Mason historians and art historians were thoughtful critics as well as excellent company. Rob Townsend, in his role as an assistant director of the American Historical Association, was a fantastic partner in my efforts to understand and shape federal policy toward oral history.

Beyond Mason, I was kept writing by the kind words and careful criticism of those who found value in the work, particularly Philip Hamburger, Jack Katz, Jonathan Knight, Roger Launius, John Mueller, Don Ritchie, Carl Schneider, Linda Shopes, and Simon Whitney. Steve Burt pushed me to write better and to think better, and he helped translate my thoughts into readable prose. I am especially grateful to everyone who commented on my Institutional Review Blog, and to those who responded to the OHRP's call for comments in late 2007. Three anonymous reviewers for the *Journal of Policy History*, and that journal's editor, Donald Critchlow, offered welcome feedback on one part of this story. The resulting article appeared as "How Talking Became Human Subjects Research: The Federal Regulation of the Social Sciences, 1965–1991," *Journal of Policy History* 21, no. 1 (2009): 3–37, copyright © 2009 Donald Critchlow and Cambridge University Press. Revised portions of that article are incorporated into chapters 2, 5, and 6. They are reprinted with permission.

To answer the questions raised by all of these colleagues, I depended on the help of archivists, researchers, and policy makers, including Mark Frankel, Lucinda Glenn, David Hays, Suzie Janicke, Richard Mandel, Silvia Mejia, Michael Murphy, and Nora Murphy, as well as those participants in this story who agreed to be interviewed and who are cited in the notes. The OHRP Web site and the Department of Health and Human Services' Freedom of Information office were excellent sources of material. Tony Lee and Gail Saliterman offered lodging and companionship during my research trips.

At the crucial point when I had to decide whether I was writing a couple of articles or a full-length book, Bob Brugger of the Johns Hopkins University Press steered me toward the latter. Along with Courtney Bond, Kathleen Capels, Juliana McCarthy, Josh Tong, two anonymous readers, and the rest of the Hopkins crew, he has been a steady source of encouragement, wisdom, and style. The index was prepared by Anne Holmes of EdIndex and was funded by the George Mason University Department of History and Art History.

Finally, I am grateful to have the support of a wonderful family. Rebecca Tushnet kept me honest, demanding evidence and clarification. Elizabeth Alexander, Rhoda Bernard, Bob Fenichel, Lisa Lerman, Sam Lerman, David Schrag, Lala Schrag, Sarah Schrag, Eve Tushnet, and Mark Tushnet offered inspiration and sometimes even child care. Leonard Schrag and Nora Schrag asked thousands of probing questions without getting anyone's permission.

I dedicate this book to my father, Philip Schrag, and to the memory of my mother, Emily Fenichel. They raised me to value inquiry, ethics, freedom, and, when necessary, dissent.

ABBREVIATIONS

ACHRE	Advisory Committee on Human Radiation Experiments, 1994–1995
ADAMHA	Alcohol, Drug Abuse, and Mental Health Administration (part of the PHS, 1973–1992, after which it was renamed)
DHEW	Department of Health, Education, and Welfare, 1953–1980
HHS	Department of Health and Human Services (DHEW's successor, 1980–present)
National Commission	National Commission for the Protection of Human Subjects of Biomedical and Behavioral Research, 1974–1978
NBAC	National Bioethics Advisory Commission, 1995–2001
NIH	National Institutes of Health (part of the PHS)
NIMH	National Institute of Mental Health (part of the PHS)
OHRP	Office for Human Research Protections (OPRR's successor; part of HHS, 2000–present)
OPRR	Office for Protection from Research Risks (part of the NIH, 1974–2000)
PHS	Public Health Service (part of DHEW, then HHS)
President's Commission	President's Commission for the Study of Ethical Problems in Medicine and Biomedical and Behavioral Research, 1978–1983

ETHICAL IMPERIALISM

INTRODUCTION

On 18 November 2004, Hunter College told Professor Bernadette McCauley to cease all her research; she was under investigation. In a letter sent by certified mail both to McCauley's home and office, two professors warned her that she had put the entire City University of New York at risk, and that her actions would be reported to the federal government.[1] What dreadful crime had McCauley committed? She had obtained the phone numbers of four nuns who had grown up in Washington Heights and who might give advice to some history students researching the neighborhood. McCauley thought she was just arranging some casual conversations.[2] To Hunter College's Committee for the Protection of Human Research Participants, however, McCauley's potential interaction with the nuns might be "research which involves human subjects." By failing to seek permission to conduct it, McCauley had committed a serious academic offense. The committee began to investigate not only McCauley's contact with the nuns, but also her use of archival documents while researching a book. McCauley thought this was absurd, and she hired a lawyer to resist the pressure. But on being told her job was at risk, she gave the committee the information it wanted. After six months, it concluded that her work was not subject to review after all.[3]

McCauley's case was extreme, but it illustrates the powers that *institutional review boards* (IRBs), such as Hunter's committee, can wield over almost every scholar in the United States who wants to interview, survey, or observe people, or to train students to do so. More and more since the late 1990s, these IRBs have claimed moral and legal authority over research in the humanities and social sciences. Researchers who wish to survey or interview people must first complete ethical training courses, then submit their proposed research for prior approval, and later accept the boards' changes to their research strategies. Researchers who do not obey the boards may lose funding or promotions, or, in the case of students, be denied degrees.

Most social science projects proceed through IRB review relatively unimpeded, but a significant number do not. Since the 1970s—and especially since the mid-1990s—IRBs have told many scholars they can only ask questions under conditions that make research difficult or impossible. These researchers are angry about the obstacles put in front of them, but they are not the only losers. While we cannot measure the research lost to ethics review, it is clear that as a result of IRB oversight, society as whole has fewer books and articles, and has fewer scholars willing to interact with other human beings. As a result, we know less about how casinos treat their employees and how doctors treat their patients. We know less about the daily workings of legislatures and strip clubs, about what it's like to be a gay Mormon or an AIDS activist. We know less about why some people become music educators and others become suicide bombers.[4] And we have ever more students—undergraduates and graduates—who are being discouraged from academic careers that depend on talking to other people. For anyone who values scholarship, IRBs are a matter of concern.

All academic research has implicit or explicit requirements, often involving committees of peers. Scholars, by the nature of their profession, agree to conduct their research within certain rules. A scholar who plagiarizes or falsifies data may not have committed a crime or a tort. Yet, if found out, he may well be disciplined, even dismissed from his job. On a more routine level, scholars must impress committees of their peers in order to have articles and books published by scholarly presses, gain tenure, and win research grants.

But the human subjects rules that IRBs enforce are strikingly different from those evaluations in two respects. First, they seek to restrain scholars at the start of their investigations. Rather than trusting scholars to behave and then punishing offenders, as is the case with rules against plagiarism or fabrication, the IRB system seeks to predict which projects might do harm. For oral historians, ethnographers, and other researchers who begin a project not knowing whom they will meet and what questions they will ask, such a prediction is often impossible. Second, while scholars seeking publication or accused of research misconduct can generally hope to be judged by experts in their fields, IRBs are never composed of researchers in a single discipline in the social sciences or humanities, and they may not have any members familiar with the ethics and methods of the scholars who come before them. Both of these anomalies can be traced to IRBs' origins as a response to abuses in medical and psychological experimentation, and to the continued dominance of ethics-oversight bodies by medical and psychological researchers and ethicists.

This is not to say that IRB review works well for medical and psychological research, either. Even in these fields, researchers and practitioners—including those supportive of the general idea of ethics review—often complain about the reality of IRB oversight.[5] Some even question the legitimacy of the system as a whole. As Carl Schneider, a member of the President's Council on Bioethics, lamented in 2008:

> A distressingly strong case can be made that the IRB system is an improvident steward of the resources it commands—that it is structured to make capricious and unwise decisions; that it delays, damages, and stops research that could save lives and relieve suffering; that it institutionalizes an impoverished ethics; that it imposes orthodoxy where there should be freedom; that it corrodes the values the First Amendment protects; that it flouts the most essential principles of due process; that it is a lawless enterprise that evades democratic accountability.[6]

Given these complaints, it is important for scholars to ask whether today's IRB system efficiently protects the subjects of medical and psychological experimentation while allowing needed research. But that is not the purpose of this book.

Instead, it focuses on the distinct questions raised by IRB review of research in the social sciences and humanities. Even the use of those terms raises tricky questions. For decades, participants in the IRB debate have argued over the classification of such academic disciplines as anthropology, economics, education, geography, history, linguistics, political science, psychology, and sociology. Those who favor broad IRB jurisdiction like to lump them all together as "behavioral science" or, more recently, as "social, behavioral, and educational research" (SBER). Presenting these various disciplines as different flavors of the same basic enterprise suggests that they are subject to a federal law covering behavioral research, and that they can all be governed by a single SBER IRB at each university. In contrast, critics of IRBs are more apt to distinguish between behavioral science and social science on the grounds that the former's reliance on lab or classroom experiments (as in much of the research in psychology, education, and behavioral economics) make it fundamentally different from disciplines that rely on surveys, interviews, observation, and, occasionally, intervention in the real world. Such a distinction has political consequences, since if survey research, for example, is a social science but not a behavioral science, then it may not be covered by federal human subjects law.

There is no neutral way of discussing these subjects, and this book reveals its allegiance by reserving the term *social science* for such fields as anthropology, geography, history, political science, and sociology. Issues raised by IRB oversight of the humanities (such as folklore) and journalism are quite close to those surrounding the social sciences. For this reason, this book occasionally uses the term social science as shorthand for the clumsier "social science, humanities, and journalism." Psychology, by contrast, is termed a *behavioral* science. And this narrative does not cover in any depth the history of IRB oversight of psychology, considering it a story separate from that of IRB oversight of social science.

Part of this distinction is institutional. Psychologists have long held prominent positions within the health agencies of the federal government and on the various commissions that shaped IRB policy. They never lacked a seat at the table. Social scientists, by contrast, were left howling outside the door. Even today, human subjects research is so central to medical and psychological research that it is easy to fill IRBs with researchers who share general methodological frameworks with the investigators they review. Such common ground is often absent when IRBs govern social research.

More important distinctions are ethical and methodological ones. In much psychological and educational research, as in medical research, the following assumptions, central to the IRB system, are generally true:

1. Researchers know more about their subjects' condition than do the subjects themselves.
2. Researchers begin their work by spelling out detailed protocols explaining what hypotheses they will test and what procedures they will employ to test those hypotheses.
3. Researchers perform experiments designed to alter subjects' physical state or behavior, rather than simply gathering information through conversation, correspondence, and observation.
4. Researchers have an ethical duty not to harm their subjects.

In much research in the social sciences and the humanities, these assumptions do not hold. Social researchers often identify a topic or a community that interests them without specifying particular hypotheses or procedures, and the people whom they study usually know much more about the consequences of any interaction than does the newly arrived researcher. While some social scientists seek to change the lives of the people they study, many are content to talk and observe. And while each social science discipline has devised its own ethical code, these codes do not always encompass the basic presumptions of medical ethics, including the duty to do no harm.

Scholars in the social sciences and humanities have reacted to IRB rules in a variety of ways. Professors like McCauley, who deliberately challenge IRBs, are few. The vast majority acquiesce—filling out forms, completing ethics training, and waiting for IRB approval. Some do so willingly, finding in the process an opportunity for feedback from scholars and administrators committed to ethical research. Others do so grudgingly, fearful of IRBs' power. Still other scholars ignore their IRBs, either because they are simply unaware of their universities' rules or because they hold the IRB in contempt and doubt its willingness to pick a fight. Others restrict their research plans to publicly available documents and datasets, thus avoiding IRB jurisdiction. And some scholars turn to professional organizations to lead them in a fight against IRB control.

Then there are those scholars who use their professional training to analyze the problem of IRB oversight. Law professors have studied the legal principles.[7] Anthropologists have studied the power relationships.[8] Communication scholars have studied the rhetoric.[9] This book, by contrast, explores the *history* of the controversy. Today's government officials, university administrators, and scholars offer rival interpretations about the meaning of laws and reports written decades ago, with little understanding of the context in which those documents were developed. Previous accounts of the history of IRB oversight of the social sciences are based on surmise and lore, rather than archival research.

Some scholars, especially supporters of IRB review, believe that regulations were developed in response to specific problems in social science research. For example, the CITI program, a training program widely used by IRBs, states that "development of the regulations to protect human subjects was driven by scandals in both biomedical and social/behavioral research."[10] This is perhaps based on Robert Broadhead's unsubstantiated 1984 claim that "IRBs were given [their] responsibilities because of a history of widespread ethical violations in both biomedical and social science research."[11] Laura Stark downplays the significance of specific scandals but still argues that "from the outset, human subjects protections were intended to regulate social and behavioral researchers."[12] Eleanor Singer and Felice Levine, who understand that current regulations are "built on a biomedical model," nevertheless assume that the 1974 regulations of the Department of Health, Education, and Welfare were prompted in part by the controversial research of Laud Humphreys and Stanley Milgram.[13]

Other scholars, dismayed by the inconsistencies of current regulations and guidance, and cognizant of their awkward application to nonbiomedical research, infer that regulators included such research by mistake. Murray Wax has claimed that survey and fieldwork research were added to the regulations "almost as an afterthought."[14] Rena Lederman writes of "the accidental character of the federal code."[15] And Robert Kerr finds evidence in the regulations of an "original purpose of protecting human subjects from Tuskegee-like abuses in the course of biomedical research."[16] The "Illinois White Paper," a recent critique of IRB review, describes the regulations as having emerged "in response to specific historical events and medical developments," not from attention to the social sciences.[17]

Because such claims about the origins of today's regime affect the daily work of scholars, it would be useful to know how accurate they are. Yet, as a 2003 report noted, "nowhere is there a comprehensive history of human participant protection in social, behavioral, and economic sciences research." While that report's chapter on the history of federal regulation of such research offers a tolerable overview of the subject, its authors concede that "we are not able to make up for the lack of a comprehensive history here."[18]

This book seeks to provide that history. It draws on previously untapped manuscript materials, as well as published sources and—ironically or not—oral histories. It asks:

- Why did federal officials decide to regulate the social sciences?
- How did they decide on the provisions now contained in federal law?
- How has the government treated research based on surveys, interviews, or observing behavior in public places?
- How have universities applied federal law, regulations, and guidance to research conducted by their faculty and students?
- How have scholars reacted to ethics review, individually and collectively?

To answer these questions, this book traces three intertwined stories that led to today's human subjects regime. The first is the story of debates about research ethics, conducted within various scholarly disciplines and by several federal bodies charged with the task of drafting ethical codes. Anthropologists, sociologists, historians, folklorists, and others have long recognized that involving living participants in their research may impose ethical burdens beyond those common to all scholarly research (such as the duty to cite one's sources). Since the mid-1960s, these scholars have debated the exact nature of their ethical obligations, and the various scholarly disciplines emerged with strikingly different conclusions about the proper relationship between researcher and research participant. But the ethicists charged with advising regulators cared little for these differences, instead basing their recommendations solely on the ethical codes of medicine and psychology.

The second story is the story of the law—the federal legislation and regulations that tell IRBs what they can and must do. Here, too, medicine and psychology set the agenda. Congressional hearings on human subjects research focused almost exclusively on medical research, and the resulting federal law requires IRB review only for "biomedical and behavioral research," not research in the social sciences or humanities. The officials who translated that law into regulations were, for the most part, health researchers themselves or employees of health agencies. They designed regulations for medical experimentation, yet most of them insisted that those regulations apply to the social sciences as well.

The third story is the application of laws and regulations by universities and their IRBs. Since their inception, IRBs have had an ambiguous relationship with federal regulators. On the one hand, the federal government claims to trust local committees to do the right thing. On the other hand, from time to time it promulgates guidance contrary to the wishes of universities and their IRBs. And, increasingly, it has enforced its regulations by suspending, or threatening to suspend, federal funding, a vital source of revenue for every research university. The result is a network of IRBs that are neither wholly independent nor wholly coordinated in their application of federal regulations, and in which neither IRBs nor regulators can be held accountable for abuses. When confronted by critics, IRBs and federal regulators can blame each other, rather than taking responsibility for their decisions.

Of these three stories, this book emphasizes the second: that of public policy, particularly on the federal level. Today, federal officials boast of the "flexibility" afforded by regulations, and they blame any abuses on university officials who have failed to take advantage of that flexibility. Such arguments ignore the long history of federal interference in university affairs. In particular, in the 1970s and 1990s regulators cracked the whip, threatening universities with the loss of millions of dollars in federal research grants if they did not comply with the regulators' preferred interpretations of nominally flexible rules. The universities, in turn, imposed strict and often inappropriate rules on social scientists. Social science organizations had to advise their members to either resist or comply. In contrast, in the 1980s and early 1990s federal officials interpreted the rules with a relatively light hand, and universities gave their faculty considerable freedom. Since other nations—particularly Australia, Canada, and the United Kingdom—used American policies as models, even scholars on the far side of the planet are affected by decisions made in and around Washington, D.C. Because U.S. federal policy makers had so much power, this book presents them as the key actors in this narrative.

At every level, this history is not as simple as either side in the IRB debate might expect. Ethicists, federal officials, and university administrators did raise sincere concerns about dangers to participants in social science research, especially about an unwarranted invasion of privacy. They also understood the objections raised by social scientists and sought ways to limit unnecessarily cumbersome review. Thus the regulation of social research was not mere accident.

On the other hand, the application of the regulations to the social sciences was far less careful than was the development of guidelines for biomedical and psychological research. In the 1970s, medical experimentation became a subject of national debate, with lengthy hearings in Congress, a new federal law covering "biomedical and behavioral research," and additional hearings before and deliberations by specially constituted commissions. Though the object was to put limits on the medical profession, medical doctors and biomedical researchers remained powerful enough to ensure that they would be consulted at every step of policy formation. As witnesses, commission members, and public officials themselves, these researchers could be certain that their concerns would be heard. Psychologists, though junior partners in the endeavor, also occupied positions of authority.

In contrast, social scientists lacked political power and so remained the objects of policy, rather than its authors. Regulators spent little time investigating the actual practices of social scientists, relying instead on generalities and hypothetical cases. Key questions about interview, survey, and observational research were confined to confidential discussions among federal officials, from which social scientists were excluded, and the regulations themselves were drafted in haste. By 1979, when the applicability of the regulations did become subject to public debate, it was too late for a fresh examination of the problem. The social sciences were the Rosencrantz and Guildenstern of human subjects regulation. Peripheral to the main action, they stumbled onstage and off, neglected or despised by the main characters and destined for a bad end.

For the most part, regulators did not ask whether the social sciences needed regulation, or even what regulations would best serve participants in social science research. Instead, they asked a narrower question: how policies developed for biomedical research could be adapted to the social sciences. The creators of the IRB system sought policies that could be applied to all kinds of research funded by all federal agencies. In search of such universals, they framed the debate over social science in terms borrowed from biomedical science—including biomedical definitions of terms like science, research, and human subjects. They ignored important differences among disparate types of research. For the sake of simplicity, regulators forced social science research into an ill-fitting biomedical model. At times they pledged that a policy specific to the social sciences was just around the corner. At other times they passed the buck, pledging that someone else would fix the problem if offered enough flexibility. But never did they do the hard work of studying the problems raised by social science research and designing a system to address those problems.

Why regulators behaved so badly is a difficult question. Policy makers themselves repeatedly claimed to have devoted more close attention to social science research than they actually had, making it hard to give full credit to their own explanations. Instead, it seems that they combined good intentions with an unwillingness to consider all the consequences of their actions. The most powerful shapers of the IRB system shared a sincere interest in the rights and welfare of participants in research, especially medical research. To protect those participants, they wrote rules that would pertain to as much research as possible. At times they deliberately included social science research, just in case, someday, they decided they needed to rein in social scientists. More frequently, their real concern was to cover all medical and psychological research, and that meant using inclusive language that encompassed activities they had no particular interest in governing.

Ethicist Albert Jonsen recognized this problem in April 1978, using the term *ethical imperialism* to describe the imposition of one field's rules onto another discipline. "I think that the amateurs in ethics suffer in imperialism," he warned. "They practice imperialism, which is constituted by establishing a rule and applying it ruthlessly. You know, wherever you go, you set your rule down. What we ought to be particularly good at by now is seeing that rules have very different meanings as situations approach or depart from a kind of a paradigm situation."[19] It is a fine metaphor, so long as one remembers historian John Seeley's claim that the British had "conquered and peopled half the world in a fit of absence of mind"; federal policy makers were just as inattentive as they wrote the rules that govern social science.[20] Today, this combination of ignorance and power—so potent on the federal level—is recapitulated in IRBs across the country. Many boards make ill-informed, arbitrary decisions while congratulating themselves on their ethical rigor.

Even supporters of IRB review of the social sciences recognize that boards and university administrations often make mistakes. The question is whether these malfunctions result from operator error or from design flaws inherent in the machine. Answering that question requires tracing the history of IRBs from their start in the 1960s up through the present day.

CHAPTER ONE

ETHICS AND COMMITTEES

Social science is not a single scholarly discipline, but a loosely defined set of approaches to studying human beings' thoughts, words, and deeds. Though many of today's social science disciplines can trace their heritage to a shared group of thinkers in the nineteenth century, by the early twentieth century they had split apart, methodologically and institutionally. The disciplines' shared and diverse heritage informed a series of ethical debates in the mid-twentieth century. In many cases, different disciplines faced common challenges, and at times they came up with common responses. But the institutional and intellectual separation of scholarly societies led them to devise ethical codes distinct from one another and distinguishable from the codes adopted by medical researchers.

The ethical debates among social scientists did not greatly shape subsequent regulations, which were devised by medical and psychological researchers and ethicists who knew and cared little about the work of social scientists. Nevertheless, it is important to note the path not taken. Just as researchers and officials were instituting one set of ethics for medical experimentation, social scientists were engaging in their own discussions, emphasizing concerns about sponsorship, deception, and confidentiality. Had regulators wished to explore the ethics of social science, by the early 1970s they would have had much to read.

Instead, the policy makers who created IRBs focused solely on medical and psychological research and spent almost no time pondering the effect of their rules on scholars doing survey, interview, and observational research. Although the rules they wrote covered social science, those rules had little relationship to the ethics of the social science disciplines.

THE ETHICS OF SOCIAL SCIENCE

Though people have been observing other people since the species began, social science as we know it is the invention of the Enlightenment. Inspired by the achievements of such natural scientists as Francis Bacon and Isaac Newton, thinkers in Scotland, France, and elsewhere hoped to use observation and statistics to understand history, politics, and behavior.[1] As late as 1890, what Americans and Europeans called social science consisted of practices that today might be regarded as "a hodgepodge of philosophy, social theory, history and hard science methods."[2] Most of these disciplines, along with more utilitarian efforts at social reform, found an institutional home in the American Social Science Association.[3]

But starting in the 1880s, German-educated American scholars began defining their research methods more narrowly and forming more specialized organizations. Between 1884 and 1905, the American Historical Association, the American Economic Association, the American Political Science Association, and the American Sociological Association all emerged to promote scholarship in more specialized fields. Anthropologists claimed a broader scope; in 1895, one anthropologist declared his discipline's ambition to "survey the whole course of the life of the [human] species, note the development of its inborn tendencies, and mark the lines along which it has been moving since the first syllables of recorded time; for this, and nothing less than this, is the bold ambition toward which aspires this crowning bough of the tree of human knowledge."[4] Given their proclivity toward physical research—collecting artifacts and measuring skulls—anthropologists found a home in the American Association for the Advancement of Science, where they remain today as Section H.[5] But they were specialized enough to want their own professional society as well. In 1902, after some debate, they founded the American Anthropological Association as a broadly inclusive organization.[6]

Institutional divisions reflected significant intellectual divergence. For example, historians and political scientists, who for a time shared the American Historical Association, split over the purpose of studying the past. Historians sought to study the past for its own sake, while political scientists sought rules and models that could inform present policy. In 1903 the latter founded the American Political Science Association, marking that discipline's separation from the field of history.[7] Likewise, the American Sociological Association was a 1905 spinoff from the American Economic Association.[8] Anthropologists had their own split in 1941, when a group formed the Society for Applied Anthropology. As its name implied, the society hoped not only to gather knowledge about human interaction, but to apply such knowledge for the benefit of administrators in government and business. By the 1950s, members of the society were distancing themselves from other anthropologists and claiming the title "social scientists."[9]

Although sometimes labeled a social science, psychology, it must be emphasized, has a distinct history. While the psychologists of the late nineteenth and early twentieth centuries employed a range of techniques for a variety of problems, many marked their discipline by the use of laboratory experiments, including precise measurements borrowed from physiology.[10] This experimental focus alienated the philosophers who had helped found the American Psychological Association in 1892, and in 1900 they departed to join the new American Philosophical Association. The remaining psychologists sought to bolster their discipline's status by emulating the methods of physical science.[11]

At some level, anthropology, economics, folklore, history, political science, psychology, and sociology all sought to explain why people do what they do, and from time to time scholars tried reuniting one or more of the disciplines. In 1949, psychologist James Grier Miller of the University of Chicago formed a Committee on the Behavioral Sciences in an effort to harness the tools of both the biological and social sciences. Throughout the 1950s, Miller continued to popularize the term behavioral science, founding a journal with that title and chairing a National Academy of Sciences committee on behavioral science. This approach to human behavior was modeled on biology, and Miller hoped to quantify behavior and develop testable hypotheses.[12] The Ford Foundation endorsed the approach, creating a Behavioral Sciences Division, though the meaning of the term varied from place to place.[13]

Other scholars resisted the behavioral sciences label, arguing that disciplinary boundaries could clarify assumptions about what questions should be asked and how they should be answered. Even as sociologists and anthropologists joined psychologists in the behavioral science movement, historians and political scientists remained aloof. In 1955, for example, political scientist Arthur Macmahon noted that while the physical scientist "probes for the suspected underlying simplicities of the physical universe," it was the role of the social scientists to "face toward diversity."[14] And in 1963, sociologist Charles Bolton objected to attempts to link psychology with sociology, arguing that the former's interest in individuals and physiological changes could distract sociologists from their study of groups, social interaction, and collective behavior.[15]

Despite their varying histories, the social sciences did share some ethical concerns in the mid-twentieth century, particularly concerning questions of sponsorship, deception, and confidentiality. The question of government sponsorship was the oldest, for anthropologists had long worried that their research was being used to further imperial domination. This issue had arisen as early as 1919, when Franz Boas denounced four fellow anthropologists for having "prostituted science by using it as a cover for their activities as spies." During World War II, scholars had cheerfully served the war effort without provoking much controversy within or without their professions, though later scholars would question the ethics of some of their wartime actions.[16] In the 1950s, the nascent Central Intelligence Agency (CIA) used anthropology as a cover for some of its officers, and some actual anthropologists worked for the agency. In other cases, the CIA channeled its money through front organizations, thus funding social scientists without their knowledge.[17]

In the 1960s, government sponsorship reemerged as a major ethical problem, and not just for anthropologists. In 1965, the Chilean press reported that the U.S. government was sponsoring social science research in Latin America with some kind of nefarious goal—perhaps a coup in Chile itself. As the full story emerged, it turned out the U.S. Army was, in fact, planning to spend millions of dollars over three or four years on Project Camelot. The project was designed not to topple governments, but to sponsor social science that might help existing governments avoid Communist insurgencies.[18] No one accused the scholars recruited for the program of active spying, but many critics fretted that had the project continued, the scholars involved would have sacrificed their intellectual integrity to the goals of the U.S. military. Similar concerns arose over the next several years as the escalating war in Vietnam led Pentagon and CIA officials to seek scholarly perspectives on various Asian countries. In 1970, antiwar activists revealed that the U.S. government had sought to recruit anthropologists for counterinsurgency work in Thailand, leading to another round of charges.[19] By that time, members of Congress frustrated by the Vietnam War in general had restricted the Pentagon's ability to fund social science projects that were not directly related to military operations.[20] In these and other projects, social scientists struggled to reconcile their dual roles as scholars devoted to the neutral pursuit of knowledge and as citizens wishing to further the cause of democracy.[21]

The controversies over Project Camelot and programs in Thailand raised the specter of spies posing as researchers. A second debate concerned the converse: researchers posing as something other than researchers or otherwise deceiving people about their true scientific goals. This debate primarily centered on deceptive psychological experiments, most famously Stanley Milgram's obedience studies of the early 1960s, in which he falsely told subjects that they were administering severe electric shocks to other people. For some subjects, the experience was upsetting, leading them to sweat, tremble, and stutter at the thought of the pain they were inflicting on strangers. Yet in a follow-up survey, 84 percent of the subjects told Milgram they were glad to have participated, though some remained angry enough to complain to Milgram or the president of his university. Milgram also faced criticism from his fellow psychologists.[22]

Social researchers rarely used such elaborate hoaxes, but on occasion they did employ concealment or deception. In 1952, a group of law faculty at the University of Chicago persuaded a federal judge to let them secretly record the deliberations of six juries. When they presented their research, an outraged Senate subcommittee held hearings, and eventually Congress made it a crime to record juries. More common were cases where researchers were visible but did not identify themselves as researchers. In the 1950s and 1960s, several sociologists and anthropologists presented themselves as something they were not—a mental patient, a religious mystic, an enlisted airman, or an alcoholic—in order to infiltrate and observe the settings where such people gathered: mental health care, a spiritual community, military training, and Alcoholics Anonymous.[23] In still other cases, researchers freely admitted that they were conducting research but lied about their exact topics. For example, Pierre van den Berghe told South African authorities that he planned to study their country's economy, when he was really interested in apartheid.[24]

Sociologists debated the ethics of such work. In a published exchange in 1967 and 1968, Kai Erikson of Yale and Norman Denzin of the University of Illinois disagreed over whether the figurative masks work by the researchers in such cases were fundamentally different from the many masks all people wear in the course of their lives. Erikson argued that it was "unethical for a sociologist to deliberately misrepresent his identity for the purpose of entering a private domain to which he is not otherwise eligible" or "to deliberately misrepresent the character of the research in which he is engaged"; Denzin was less sure, and neither could say that the profession agreed with him.[25] For anthropologists, this deception was of less concern—many looked and acted so different from the people they studied that disguising themselves would have been hopeless.[26]

A third debate concerned the researcher's duty to keep confidential the names and identities of the individuals and communities they studied. In the early twentieth century, the value of anonymity was more often assumed than articulated. For example, a set of 1938 instructions for a first-time social researcher explained that the researcher's "unwonted habit of prying into the personal, intimate affairs of the informant . . . can only be justified on the basis of the scientific purpose of the research, the confidence in which the material is held, and the anonymity of the individual interview as it appears in the final article or monograph."[27] By the 1940s, sociologists and anthropologists had taken to inventing pseudonyms for whole towns.[28]

Promising confidentiality was easier than delivering it. In 1958, Arthur Vidich and Joseph Bensman published *Small Town in Mass Society*, based on fieldwork Vidich had done in a community not far from Cornell University. The director of the project had promised residents that no individuals would be identified in print, and Vidich and Bensman did use pseudonyms when referring to individuals. But, they later explained, "we believed that it was impossible to discuss leadership without discussing leaders, politics without mentioning politicians, education without treatment of educators, and religion without ministers."[29] As a result, someone familiar with the town reading the final text could easily identify some individuals based on their roles (e.g., school principal), along with possibly derogatory information about them. Because the project had evolved over several years, with some researchers making statements or proclaiming policies rejected by other members of the research team, it remained murky whether Vidich had broken any promises which he was obliged to honor. But several scholars—including the director of the project—felt that he had acted badly.[30]

Elsewhere, researchers disguised the names of towns and people, only to see others disclose them. As anthropologist Harold Orlans noted in 1967, "it is a popular pastime of academic cognoscenti to disclose 'anonymous' towns and authors," and "it is standard form for book reviewers to reveal the name of an 'anonymous' community."[31] A typical example was anthropologist Oscar Lewis's book, *The Children of Sánchez*, based in large part on verbatim interviews with a poor family in Mexico City. Though Lewis had done what he could to disguise his informants' identities, after the Spanish translation was published, a newspaper claimed to have identified the family after a twenty-seven-day investigation. (Fortunately for the family, the paper did not print the names.)[32]

By 1967, Lee Rainwater and David Pittman had questioned sociologists' "automatic assumption that we offer to maintain the privacy of our informants," pointing out the vulnerability of research notes to a subpoena.[33] Such concerns were realized in 1972 when political scientist Samuel Popkin chose jail rather than reveal his research data to a grand jury. The American Political Science Association applauded his principled stand, then joined with several other social science organizations to study the problem of confidentiality in research. The research team found that "social science investigators apparently routinely extend promises of confidentiality without always understanding the legal, ethical, and other implications of those promises." It called for scholars to develop better standards for research when confidentiality was involved, and for Congress and state legislatures to pass laws protecting research data from subpoena.[34]

Concerns about confidentiality—along with worries about sponsorship and deception—reflected social scientists' wishes to be honest in their dealings with others and to keep the promises they made. To some extent these concerns paralleled those of medical ethics. In their 1953 call for "an explicit code of ethics" for social scientists, Joseph Fichter and William Kolb raised the specter of sociologists behaving like Nazi doctors—abusing the rights of individual humans in their overzealous pursuit of truth.[35] But social scientists also took care to distinguish their practices from those in medicine. First, some questioned the need for informed consent, a medical term not used by anthropologists until it was imposed by federal rule makers.[36] As anthropologist Margaret Mead argued in 1969, "voluntary participation . . . in a collective enterprise" might be a more appropriate standard than informed consent.[37] For their part, sociologists wondered what it would mean to get informed consent from an entire business or school.[38]

Second, many social scientists rejected the physician's pledge to do no harm. Rather, they advocated following the truth, even into unpleasant places. Even Fichter and Kolb, who called for social scientists to spare human feelings whenever possible, were happy to see researchers report "in full detail" the activities of gangsters and Klansmen, and other researchers accepted a longer list of targets.[39] Vidich, defending his work, argued, "One can't gear social science writing to the expected reactions of any audience, and, if one does, the writing quickly degenerates into dishonesty, all objectivity in the sense that one can speak of objectivity in the social sciences is lost."[40] And in 1967, Rainwater and Pittman were asked by colleagues whether it was ethically acceptable to report the behavior of poor, African American residents of public housing: "If one describes in full and honest detail behavior which the public will regard as immoral, degraded, deviant, and criminal, will not the effect be to damage the very people we hope our study will eventually help?" They replied with an embrace of an unpleasant truth:

> If you believe that in the long run truth makes men freer and more autonomous, then you are willing to run the risk that some people will use the facts you turn up and the interpretations you make to fight a rear guard action. If you don't believe this, if you believe instead that truth may or may not free men depending on the situation, even in the long run, then perhaps it is better to avoid these kinds of re-

> search subjects. We say perhaps it is better, because it seems to us that a watered-down set of findings would violate other ethical standards, and would have little chance of providing practical guides to action; thus it would hardly be worth the expenditure of one's time and someone else's money.

While they expected that this principle would mainly apply to the study of groups, they also envisioned ethical criticism of individuals' "publicly accountable behavior." Indeed, they maintained, "one of the functions of our discipline, along with those of political science, history, economics, journalism, and intellectual pursuits generally, is to further public accountability in a society whose complexity makes it easier for people to avoid their responsibilities."[41] Mead agreed: "The research worker may feel that the exposure of some evil transcends his obligation to the image of his science that will be formed by angry resentment of his disclosures."[42] Sociologist Howard Becker put it more succinctly: "A good study . . . will make somebody angry."[43]

ETHICAL CODES

Troubled by these debates, the various professional societies either adopted statements or codes of ethics for the first time in the 1960s, or expanded the scanty codes they had adopted in the 1940s. In 1963, the Society for Applied Anthropology deliberately avoided the term code, but it did adopt a "statement on ethics," which pledged that the applied anthropologist owes "respect for [the] dignity and general well-being" of his fellow men, and that

> he may not recommend any course of action on behalf of his client's interests, when the lives, well-being, dignity, and self-respect of others are likely to be adversely affected, without adequate provisions being made to insure that there will be a minimum of such effect and that the net effect will in the long run be more beneficial than if no action were taken at all.

Significantly, this warning applied to the researcher's recommendations for action, not the research itself. The statement also warned the researcher that he "must take the greatest care to protect his informants, especially in the aspects of confidence which his informants may not be able to stipulate for themselves."[44] A 1974 revision extended the researcher's obligation to the community, promising a community "respect for its dignity, integrity, and internal variability," but it also noted that "it may not be possible to serve the interests of all segments of the community at the same time that we serve the interests of the contracting agency."[45]

The American Anthropological Association (AAA) was slower to adopt a statement. Following the controversy over Project Camelot, attendees at the association's 1966 meeting called for an exploration of the relationship between anthropologists and their sponsors, both private and public.[46] The result was the 1967 "Statement on Problems of Anthropological Research," which directed American universities and anthropologists to stick to "their normal functions of teaching, research, and public service" and to "avoid both involvement in clandestine intelligence activities and the use of the name anthropology, or the title anthropologist, as a cover for intelligence activities." The statement also admonished anthropologists to protect "the personal privacy of those being studied and assisting in their research," and to keep in mind that "constraint, deception, and secrecy have no place in science."[47]

In 1971, following the controversy over counterinsurgency research in Thailand, the AAA adopted a far more sweeping statement on the "Principles of Professional Responsibility." While clearly concerned with clandestine work for governments, the new statement declared much broader ideals, applicable to a wider range of work:

> In research, an anthropologist's paramount responsibility is to those he studies. When there is a conflict of interest, these individuals must come first. The anthropologist must do everything within his power to protect their physical, social and psychological welfare and to honor their dignity and privacy . . .
>
> Informants have a right to remain anonymous. This right should be respected both where it has been promised explicitly and where no clear understanding to the contrary has been reached. These strictures apply to the collection of data by means of cameras, tape recorders, and other data-gathering devices, as well as to data collected in face-to-face interviews or in participant observation. Those being studied should understand the capacities of such devices; they should be free to reject them if they wish; and if they accept them, the results obtained should be consonant with the informant's right to welfare, dignity and privacy . . .
>
> There is an obligation to reflect on the foreseeable repercussions of research and publication on the general population being studied.[48]

Not all anthropologists were comfortable with the notion that a social scientist must do no harm. Dutch scholar A. J. F. Köbben applauded the American anthropologists who had targeted their research toward the defeat of the Axis in World War II and, more cautiously, van den Berghe's decision to deceive the apartheid government of South Africa.[49] Sociologist Myron Glazer found the code's absolutes "both impractical and undesirable. While making good copy to salve the conscience of legitimately troubled field workers, it can only further cloud the controversy about appropriate behavior."[50]

Meanwhile, the American Sociological Association was working toward its own code. In 1961 the association appointed its first committee on professional ethics, and in 1967 a successor committee was given the task of preparing a statement on research ethics. In 1969 the association's council approved a code with oddly contradictory precepts. The code stated that "all research should avoid causing personal harm to subjects used in research." But it continued on to state that "provided that he respects the assurances he has given his subjects, the sociologist has no obligation to withhold information of misconduct of individuals or organizations."[51] The code also avoided any mention of informed consent, perhaps to accommodate members who engaged in observational research.[52]

For political scientists, the overwhelming concern was to draw some lines between scholars' roles as objective observers and their other positions as grantees or as participants in the political process. In 1966 the news of Project Camelot and allegations of covert CIA sponsorship led the American Political Science Association to form a committee on professional standards and responsibilities. After some initial investigations, in the summer of 1968 the committee issued its final report, which dealt extensively with the problem of maintaining intellectual freedom and integrity while accepting money from an outside sponsor—whether the CIA, another federal agency, or anyone else—and it warned academic political scientists to "avoid any deception or misrepresentation concerning his personal involvement or the involvement of respondents or subjects, or use research as a cover for intelligence work." Beyond that, however, the report said surprisingly little about ethical obligations to survey and interview respondents. The report did raise questions about informed consent and confidentiality in research as seemingly innocuous as interviewing, but it did not propose any guidelines.[53]

Historians wrestled with some of the same issues. Since the time of Thucydides, historians had been interviewing witnesses to recent events, and in the 1940s they began using newly available recording equipment to record and transcribe verbatim interviews, a practice that took the name *oral history*. While most interview projects remained uncontroversial, the experience of William Manchester spooked some historians. After choosing Manchester to write an account of John F. Kennedy's assassination and then favoring him with interviews, Robert and Jacqueline Kennedy demanded editorial changes and ultimately tried to block publication of Manchester's work. The episode highlighted the fact that the story the historian seeks to tell may not be one the narrator wishes told.[54] In response, historian Willa Baum suggested the newly created Oral History Association needed a code of ethics promising that "the interviewer will not exploit his relationship to the narrator nor knowledge of the interview material to the detriment of the narrator."[55] But when the association did adopt its first statement of goals and guidelines in November 1968, it put no such burdens on the interviewer. So long as the interviewer and narrator agreed about how the recording and transcript would be prepared and used, the historian had no special obligation to safeguard the narrator's well-being.[56] In 1975, however, a proposed revision added the obligation that "interviewers should guard against possible social injury to or exploitation of interviewees and should conduct interviews with respect for human dignity," and this was adopted in 1979.[57]

The American Psychological Association (APA) had adopted an ethical code in 1953, but that had focused on misbehavior by clinical psychologists, rather than researchers. In 1970 the APA decided to supplement that code with one devoted to research.[58] In formulating its first code, the association had rejected what it termed the "armchair approach" of allowing a committee to draft a code on its own in favor of a "research approach," which would solicit from APA members "descriptions of actual situations which required ethical decisions . . . accompanied by a statement of the decision actually made and the author's judgment as to the ethical soundness of the decision."[59] Now the association followed the same procedure for a research code, eventually receiving about five thousand descriptions of ethically challenging situations that its members had faced. After drafting a set of proposed principles, the association's ethics committee got feedback from about two hundred groups and individuals. Only when it had devised a code acceptable to a broad consensus did the association adopt anything. Then, in 1973, the association published the code as a 104-page book, complete with brief descriptions of many of the incidents which had been collected in the course of the code's creation.[60]

All told, the 1960s and early 1970s were a remarkably fruitful time for social and behavioral scientists eager to debate and codify the ethics of their disciplines. But this did not mean that they reached any consensus on how researchers should behave. A 1975 analysis by sociologist Paul Davidson Reynolds of twenty-four codes of ethics—American and European—found that "no one principle or statement was present in every code of ethics."[61] And even within a discipline, no code could foresee every case or give clear answers about the ethical validity of every project.[62] As sociologist Howard Becker pointed out, the vagueness in some codes reflected a lack of consensus about what was and was not ethically acceptable research.[63] That lack of consensus was best illustrated in what remains perhaps the most controversial work of social science research in American history.

TEAROOM TRADE

In June 1965, 34-year-old Laud Humphreys entered the doctoral program in sociology at Washington University in St. Louis after a brief and tempestuous career as an Episcopal priest. A closeted (and married) gay man, he decided to study the St. Louis gay community, and his adviser, Lee Rainwater, encouraged him to look at so-called tearooms, places where men gathered for quick, anonymous sex.[64] Humphreys took the advice, and he got funding for the work from the National Institute of Mental Health (NIMH), part of the U.S. Public Health Service.[65]

Starting in the spring of 1966, Humphreys began frequenting public restrooms in St. Louis's Forest Park and observing men engaged in oral sex, generally without speaking to them. Rather than identify himself as a sociologist—which he assumed would have stopped the activity he was trying to observe—Humphreys presented himself as a participant.[66] In his published work, he claimed that he had served merely as a lookout, warning the men if a stranger was approaching, but his biographers conclude that Humphreys more likely participated in the sex himself.[67]

To learn more about the men whose behavior he had witnessed, Humphreys disclosed what he called his "real purpose for being in the tearooms" to twelve men who seemed willing to talk. These he interviewed at length. But since their very willingness to be interviewed made them unrepresentative of the tearoom visitors, Humphreys devised an elaborate research scheme that would allow him to interview a broader sample. He noted the license plate numbers of the cars of men he had seen engaged in sex. Then he told a campus police officer that he was involved in "market research" and thus got access to tables listing the owners of those cars.[68] Next he added the names of a hundred of the men to a list of people to be surveyed for a social health project. Having waited a year since observing the men, and having been "careful to change [his] appearance, dress, and automobile from the days when [he] had passed as deviant," Humphreys now presented himself as a researcher on the survey project. He asked the men about their marriages, families, jobs, and social lives, including their sex lives at home, and they responded with a range of answers, leading Humphreys to conclude that "like other next door neighbors, the participants in tearoom sex are of no one type." Humphreys himself understood the risks of his research. He faulted another sociologist for including the names of gay bars and their patrons in his master's thesis. But he did not think that ethics demanded avoidance of touchy subjects. As he later wrote, "I believe that preventing harm to his respondents should be the *primary* interest of the scientist. We are not, however, protecting a harrassed [*sic*] population of deviants by refusing to look at them."[69]

Not everyone agreed. After Humphreys first published portions of his work in a January 1970 article, *Washington Post* columnist Nicholas von Hoffman complained that "no information is valuable enough to obtain by nipping away at personal liberty, and that is true no matter who's doing the gnawing."[70] Not long afterwards, Humphreys published the complete study as *Tearoom Trade: Impersonal Sex in Public Places* and won even more attention, positive and negative.[71] Donald Warwick—trained as a social psychologist but teaching as a sociologist—condemned Humphreys for using "deception, misrepresentation, and manipulation" in his work, though Warwick did not define these terms and relied on guesswork about what Humphreys had told the men he studied. He went on to argue that Humphreys's ethics would justify Nazi medical experiments and torture.[72] Humphreys's advisors, Irving Horowitz and Lee Rainwater, defended him. They rejected an "ethic of full disclosure," noting that "to assume that the investigator must share all his knowledge with those being investigated also assumes a common universe of discourse very rarely found in any kind of research, much less the kind involving sexual deviance."[73]

Ever since its publication, IRB advocates have used *Tearoom Trade* as evidence for the need for IRB review of social research. But it makes a bad exemplar, for two reasons. First, the sort of deliberately covert and deceptive social research conducted by Humphreys was, and remains, quite rare. In 1965 and 1966, just when Humphreys was beginning his research, the two top sociological journals published 136 articles of substantive research (as opposed to methodology or theory), of which almost two-thirds relied on interviews or surveys conducted by the author or other researchers.[74] Only six relied on any participant observation, and those presumably included studies in which the researcher was quite open about his identity and the nature of his work. Covert observation probably accounted for less than one out of a hundred sociology projects, and some of Humphreys's stratagems—such as deliberately altering his appearance so he would not be recognized from one encounter to the next—may have been unique. Thus it has always been a reach to suggest that to guard against the hazards of Humphrey's combination of covert observation and misleading interviews, committees must review every interaction between a social scientist and another person.

Second, the IRB system is premised on the idea that a committee of researchers will have better moral judgment than one or two scientists, who may be so excited by the prospects of discovery that they lose sight of the humanity of their subjects. As critics have noted, this premise is doubtful even in medical research—the researchers in some of the most abusive projects did their work in plain view of their professional peers.[75] But IRBs might have intervened in cases where, it turned out in retrospect, researchers had strayed far from the accepted ethical consensus of their disciplines.

Tearoom Trade was different. Though Humphreys did not get IRB approval, he did plan his work in collaboration with his dissertation committee—an eminent group of sociologists, some of whom had published essays on the ethics of research.[76] He also discussed his work with Kai Erikson, who later took a hard line against misrepresentation.[77] He had tried to follow the few ethical precepts that had been taught him in graduate school.[78] And when the work was complete, Humphreys won a major award, along with rapid promotion and multiple job offers.[79] Those who considered his work an atrocity were up against not a single rogue researcher, but a good portion of an entire academic discipline. As Warwick, one of Humphreys's harshest critics, conceded, it was hard to condemn a project as unethical when social scientists lacked "well-developed and generally acceptable ethical standards for judging serious research."[80]

Indeed, the critics disagreed among themselves about which elements of Humphreys's complex project design had crossed the line into abuse. Myron Glazer, for example, had no complaints about Humphreys's observations in the restrooms, but felt that his deceptive interviews "make him a poor model for others to emulate without the most painful self-scrutiny."[81] After reading that assessment, Humphreys himself agreed that while he remained untroubled by his observations in the tearooms themselves, he should not have interviewed tearoom participants without identifying himself and his project.[82] The problems with Humphreys's work stemmed not from a lack of oversight, or from a scofflaw's flouting established ethics, but rather from the novelty of his questions and methods. By taking his critics seriously, Humphreys became an earnest participant in the hard work of defining the ethics of sociology.

More than anything, then, *Tearoom Trade* suggested how fluid and unresolved were the ethical questions posed by social research. In their debates over ethical codes in the mid-1960s, social scientists had shown themselves eager to find ways to promote ethical research. They had not reached consensus, but they had engaged in vigorous and informed debate. Given time, each discipline might yet agree on a set of rules, then figure out the best way to train researchers to follow them. Had policy makers wished to speed that process, they could have devoted time and resources to an investigation of social science. Instead, they merely imposed a system designed for medical research.

THE MEDICAL ORIGINS OF IRBs

IRBs were created in response to the vast expansion of medical research that followed World War II. In 1944, Congress passed the Public Health Service Act, greatly expanding the National Institutes of Health (NIH) and their parent, the Public Health Service (PHS). Between 1947 and 1957, the NIH's research grant program grew from $4 million to more than $100 million, and the total NIH budget grew from $8 million in 1947 to more than $1 billion by 1966. The same legislation authorized the NIH to open its Clinical Center, a hospital built specifically to provide its researchers with people—some of them not even sick—on whom to experiment.[83]

Medical experimentation promised long-term scientific advances, but it also carried physical and ethical risks. At the close of World War II, Americans were shocked to learn that German scientists and doctors, heirs to one of the world's great scientific traditions, had maimed and murdered concentration camp prisoners in the course of experiments on physiology and medicine.[84] Closer to home, even Jonas Salk's polio vaccine—one of the greatest public health triumphs of the twentieth century—had its victims: children infected by improperly produced vaccine. Since the NIH had taken responsibility for assuring the vaccine's safety, the infections discredited the agency, and its top officials were purged. The institutes were then put under the charge of James Shannon, one of the few officials to have urged caution in the vaccine testing.[85] Other researchers, less prominent than Salk, seemed willing to sacrifice human rights for science. In the late 1950s, a researcher fed hepatitis viruses to children admitted to a school for the mentally disabled. Though he obtained their parents' consent, and reasoned that the children were bound to contract hepatitis anyway, later critics accused him of treating the children as guinea pigs.[86] In 1963, a highly respected cancer researcher—using NIH funds—injected cancer cells into twenty-two patients of the Jewish Chronic Disease Hospital in Brooklyn. Having satisfied himself that the procedure was perfectly safe and that the word cancer would unnecessarily trouble the patients, he neither explained the experiment nor sought patient consent.[87]

In 1966, Harvard medical professor Henry Beecher published an influential article in the *New England Journal of Medicine* cataloguing these and twenty other "examples of unethical or questionably ethical studies." Beyond such episodes, NIH director Shannon was troubled by a more general sense that medical research was shifting from a process of observation to one of experimentation involving potentially dangerous medication and surgery.[88] In early 1964, well before Beecher's article appeared, the NIH had appointed an internal study group to investigate the ethics of clinical research.[89]

Searching for a system of safeguards, the group looked to the NIH's own Clinical Center. Since its opening in 1953, the center had required that risky studies there be approved by an NIH medical review committee. They also required the written consent of participating patients, who were considered "member[s] of the research team."[90] In 1965 the National Advisory Health Council—the NIH's advisory board—recommended that a comparable system be applied to the NIH's extramural grants program as well, and in February 1966 Surgeon General William Stewart announced that recipients of Public Health Service research grants could receive money "only if the judgment of the investigator is subject to prior review by his institutional associates to assure an independent determination of the protection of the rights and welfare of the individual or individuals involved."[91] As a July revision explained, grantee institutions would be required to file *assurances* with the PHS, providing "explicit information on the policy and procedure . . . for review and decision on the propriety of plans of research involving human subjects."[92]

This initial pronouncement clearly focused on medical research—one of the tasks of the reviewers was to determine the "potential medical benefits of the investigation."[93] But the NIH's behavioral sciences section also offered grants in psychology and psychiatry, as well as supporting some work in anthropology and sociology, which were termed social sciences.[94] Would these fields be covered too? As Dael Wolfle, a National Advisory Health Council member who coauthored the 1965 recommendation, later noted, "It was most assuredly not our intent that the regulation we recommended . . . be extended to research based upon survey, questionnaire, or record materials. This type of research does not involve the kinds of harm that may sometimes result from biomedical studies or other research that actually intrudes upon the subjects involved."[95] But James Shannon, the NIH director, had his own ideas. As he later recalled, "It's not the scientist who puts a needle in the bloodstream who causes the trouble. It's the behavioral scientist who probes into the sex life of an insecure person who really raises hell."[96]

Psychological probes of sex lives did, in fact, raise hell at the congressional level. At some point in the late 1950s or early 1960s, Congressman Cornelius E. Gallagher was working late in his office when he encountered a distraught woman demanding to see a congressman—any congressman. He calmed her down enough to hear her story and learned that the woman's daughter, a teenage high school graduate, had applied for a clerical position with the federal government. Though the position required no security clearance, the daughter had been subjected to a polygraph examination in which the examiner—himself only 21 and barely trained—demanded to know the details of the girl's sex life and accused her of lesbianism.[97] Gallagher was outraged enough to pursue the matter, and in 1964 he persuaded the House Government Operations Committee to hold hearings on the use of the polygraph in federal agencies.[98]

To Gallagher's surprise, the publicity surrounding the polygraph investigation led thousands of people to write to him about other federal actions they considered to be invasive of privacy.[99] Gallagher was particularly concerned about the use of psychological personality tests, such as the Minnesota Multiphasic Personality Inventory (MMPI), on federal employees and job applicants.[100] He was outraged that federal employees and applicants were being asked about their sexual preferences and religious beliefs as part of tests he considered unscientific. In June 1965, Gallagher held hearings to investigate what he called "a number of invasion-of-privacy matters," including "psychological testing of Federal employees and job applicants, electronic eavesdropping, mail covers, trash snooping, peepholes in Government buildings, the farm census questionnaire, and whether confidentiality is properly guarded in income-tax returns and Federal investigative and employment files."[101] Most of these subjects—such as federal wiretapping—had absolutely nothing to do with scholarly research. But Gallagher's subcommittee did look at three investigations that bore some resemblance to the work of social scientists.

The first two concerned the proper use of questionnaires by the federal government itself. The subcommittee devoted long bouts of questioning to the MMPI, and it succeeded in winning promises from the Civil Service Commission, the State Department, and other agencies that they would greatly restrict the use of such personality tests.[102] A second round of questioning involved the farm census, which had asked each farmer to report the income of everyone in his household, including boarders and lodgers. When Gallagher pointed out that the system required lodgers to reveal their finances to their landlords and employers, census officials conceded the problem and promised to do better in the future.[103] Though both of these examinations concerned questionnaires, they were clearly focused on the actions of federal officials, rather than on scholarly researchers.

The subcommittee also looked at "the use of personality tests and questionnaires in federally funded research activities," especially those involving schoolchildren.[104] This line of questioning came a bit closer than the others to the concern of social scientists, in that it involved research conducted by universities and colleges rather than by the federal government itself. But once again, the subcommittee's questions focused on personality tests, rather than any nonpsychological research. Likewise, the subcommittee called on the executive director of the American Psychological Association to defend his profession, but it did not summon an anthropologist, a historian, or a sociologist to do the same.

The subcommittee touched on nonpsychological research only briefly and in the most oblique ways. Norman Cornish, the subcommittee's chief investigator, noted that federal agencies sometimes excused their behavior on the grounds "that they are conducting some sort of study on human behavior and because it is a purely scientific purpose, they somehow feel this exempts them from the charge they are invading privacy." His witness, political scientist William Beaney of Princeton, replied that "I see no reason why the social scientist should be exempt from the requirements to behave decently toward his fellow man." But he did not define social science (he may have meant psychology), and when offered the chance to recommend specific measures, Beaney suggested limits on the questions put to federal job applicants and welfare recipients, rather than restrictions on scholarly researchers.[105] Congressman Frank Horton noted his objections to a federally financed study that falsely promised participants that their answers to a series of intimate questions would be kept from the researcher interviewing them for the same study.[106]

Yet IRB policy would develop around testimony that had no direct connection to social science. Under a 1942 law, the Bureau of the Budget's Office of Statistical Standards was responsible for reviewing any proposal to ask identical questions of ten or more members of the public, which covered everything from tax forms to applications for federal employment. In describing the types of federal questioning that could raise privacy issues, the Bureau's Edward Crowder mentioned "research in the field of personality and mental health, best illustrated by the research projects of the Public Health Service," which involved "questions of a personal or intimate nature." But Crowder assured the subcommittee that "questions of a personal or intimate nature often are involved, but participation is entirely voluntary, and it has been our view that the issue of invasion of privacy does not arise."[107] That seemed to satisfy the congressmen, who asked no questions about such research.

Nevertheless, near the end of the investigation, Gallagher wrote to the surgeon general on behalf of his three-member subcommittee:

> One of our primary concerns has been the use of personality tests, inventories, and questionnaires in research projects financed by grants and contracts under the Federal Government. Many of these tests, inventories and questionnaires ask our citizens to answer intimate questions about their family life, sex experience, religious views, personal values and other subjects normally regarded as solely the private business of the individual . . . It is our belief that the sponsoring agencies should adopt effective policies and guidelines to make certain that the protection of individual privacy is a matter of paramount concern and that the testing is without compulsion.

He emphasized his wish to see that "tests, inventories, and questionnaires are strictly voluntary," that investigations involving children below the college level obtain parental consent, and that information gathered by personality testing not be used for any other purpose. Beyond that, the letter did not specify any particular policies to be adopted.[108]

As we shall see, in subsequent years PHS officials argued that Gallagher's letter had prompted the service to impose institutional review on the social sciences. That seems doubtful. For one thing, Gallagher's three-man, temporary subcommittee had very little power. It did not, for example, even attempt to sponsor legislation. When agency heads resisted its recommendations, the subcommittee did not insist.[109] And even had the surgeon general wanted to obey Gallagher's wishes exactly, he would have drafted a policy dealing specifically with personality inventories, the focus of Gallahger's hearing and his letter. Clearly Gallagher was mostly concerned about psychology. When asked decades later about the kinds of scholarly research that concerned him, he mentioned psychological research by such controversial figures as Henry Murray and Arnold Hutschnecker.[110] His subcommittee had taken no testimony on the general topics of interview, survey, or observational research, and his letter mentioned neither such research nor the disciplines of anthropology, political science, or sociology. Nor, for that matter, had Gallagher demanded institutional review. As he later explained, "we did not foresee that an entirely new bureaucratic structure would be superimposed between, say, the government granting the grant and the researcher at the end, trying to do whatever it was he was trying to do. What came out of that was unintended consequences."[111] Had the PHS promised only to keep a closer eye on psychological research, there is little doubt that Gallagher would have been satisfied.

Indeed, the Public Health Service's initial response was to impose institutional review only on those methods identified by Gallagher as potentially problematic. In November 1965, Surgeon General William Stewart (the head of the PHS) assured the congressman that the PHS's "policy is one of endorsing, as guidelines in the conduct of research, the principle that participation in research projects involving personality tests, inventories and questionnaires is voluntary and, in those cases involving students below the college level, that the rights and responsibilities of the parents must be respected."[112]

Three months later, on 8 February 1966, the PHS issued Policy and Procedure Order (PPO) 129, specifying that any grant application for "clinical research and investigation involving human beings" must indicate

> prior review of the judgment of the principal investigator or program director by a committee of his institutional associates. This review should assure an independent determination: (1) of the rights and welfare of the individual or individuals involved, (2) of the appropriateness of the methods used to secure informed consent, and (3) of the risks and potential medical benefits of the investigation.[113]

This policy tracked the recommendations of the National Advisory Health Council, as indicated by the emphasis on potential medical benefits. So did a form letter drafted by the NIH, to be sent when grant applicants had submitted potentially troubling applications, reminding them of the need for a prior review process.

Two other form letters, by contrast, emphasized Gallagher's concerns but did not stress the need for prior review. One, for proposals including "the administration of personality tests, inventories or questionnaires," asked for a letter describing "the manner in which the rights and welfare of the subjects are assured. e.g., how their informed consent is obtained or why this consent is deemed unnecessary or undesirable in the particular instance." And the second, for projects involving "investigational procedures administered to children below the college level," asked for the rights of parents and guardians to be respected, including again the question of informed consent. These last two letters did not specify a review committee as a requirement for personality research or for research with children.[114] Thus, as late as February 1966, the Public Health Service's proposals for controlling medical experimentation remained separate from its safeguards for the use of personality tests and research on children. Soon, however, they would become elements of a single policy, with serious implications for all social scientists.

The early 1960s were a time of vigorous debate, in which a range of scholars and public officials grappled with questions of research ethics. The types of research they examined were as varied as experimental injections of viruses and counter-insurgency studies in South America. The ethical principles ranged from a dedication to the truth to the assertion that truth must be subordinated to the welfare of the people being studied. More specific concerns involved everything from the exposure of a farm worker's finances to the practicality of disguising an entire town. And potential mechanisms for assuring ethical research might be anything from prior review of research protocols, to the adoption of ethical codes, to the sort of published criticism that greeted *Tearoom Trade.*

While all of these debates concerned some forms of research, few observers could have predicted that they would come together as a single policy. Instead, the most promising efforts focused on specific challenges—such as the relationship between anthropologists and government agencies—rather than trying to solve all problems at once, and with a single tool. But questions of medical experimentation, congressional interest, and social science ethics would soon become entangled. And rather than devoting equal attention to each realm of inquiry, policy makers fixed their attention on medical and psychological experimentation as the disciplines whose procedures—and abuses—would inform efforts to control the conduct of all manner of research.

CHAPTER TWO

THE SPREAD OF INSTITUTIONAL REVIEW

As late as the summer of 1966, ethical debates in the social sciences and Public Health Service requirements for prior review of federally funded research seemed to have nothing to do with each other. Over the next decade, however, federal officials applied the requirements ever more broadly, so that by the early 1970s anthropologists, sociologists, and other social scientists began hearing from university administrators that they would have to get their research plans approved by ethics committees. For the most part, those writing the new rules ignored social scientists, their ethics, and their methods. This was particularly true in the congressional debates that produced the National Research Act of 1974, the federal statute that governs IRBs to this day. Less frequently, federal officials did consult social scientists, only to ignore the recommendations they received in reply. By 1977 they had begun a pattern that would persist for decades—writing policies for social scientists without meaningful input from social scientists. And social scientists had begun their own pattern of debating whether to work within the system or seek exclusion from it.

WHAT IS BEHAVIORAL SCIENCE?

The first expansion of IRB review took place within the Department of Health, Education, and Welfare (DHEW), the parent of the National Institutes of Health and the Public Health Service. Between 1966 and 1971, officials expanded the requirement of institutional review from the medical research for which it was originally designed to a much broader range of psychological, anthropological, and sociological research sponsored by the department, and they even encouraged it for projects not sponsored by DHEW but taking place in universities that accepted department funds. As these policies were being shaped, social scientists did get some opportunity to express their views, and, by and large, they objected to the review requirement. But policy-making power in this area remained in the hands of the medical and psychological researchers at the NIH, and they largely ignored the social scientists' complaints.

Two psychologists—Mordecai Gordon of the NIH's Division of Research Grants and Eli Rubinstein, associate director for extramural activities for the National Institute of Mental Health (NIMH)—seem to have been particularly influential in shaping early policies toward behavioral research. Interviewed in 1971, Gordon cited Gallagher's investigation as just one of four factors influencing the spread of the requirement, along with Gordon's own thinking, the opinions of his NIH colleagues, and letters from institutions asking whether the surgeon general had meant to include behavioral science in the PHS's policy. Though Gordon cited Gallagher's letter as the "most forceful" factor, that alone could not have steered policy in a direction opposite to the agency's wishes. More likely, Gordon and Rubinstein welcomed Gallagher's letter as a ratification of their own concerns.[1] In 1966 they took the first steps toward imposing the institutional review specified by the PHS's February order (PPO 129) on social science in general.

The NIH first publicly broached the issue of ethics review of social science in June 1966, when it sponsored a conference on ethics that brought together anthropologists, psychologists, sociologists, and other scholars, many of them members of the NIH's own Study Section in the Behavioral Sciences. The assembled social scientists acknowledged such potential dangers as psychological harms and the invasion of privacy. But, as sociologist Gresham Sykes reported, even participants who wanted clearer ethical standards had "serious reservations" about the proposed application of the PHS's new policy toward social science research. Sykes, a former executive officer of the American Sociological Association, explained their reservations in terms that would echo for decades:

> There are the dangers that some institutions may be over-zealous to insure the strictest possible interpretation, that review committees might represent such a variety of intellectual fields that they would be unwieldy and incapable of reasonable judgment in specialized areas, and that faculty factions might subvert the purpose of review in the jealous pursuit of particular interests. There is also the danger that an institutional review committee might become a mere rubber stamp, giving the appearance of a solution, rather than the substance, for a serious problem of growing complexity which requires continuing discussion. Effective responsibility cannot be equated with a signature on a piece of paper.[2]

Queried around the same time by the NIH, anthropologist Margaret Mead found the whole idea absurd. "Anthropological research does not have subjects," she wrote. "We work with informants in an atmosphere of trust and mutual respect."[3]

In the late summer and fall, two social science organizations took up the question, and both objected strongly to applying the February 1966 policy to social research. The American Sociological Association complained that "the administrative apparatus required appears far too weighty and rigid for rational use in the large majority of cases in the behavioral sciences." It also warned that "a local committee may be incompetent or biased, and may threaten the initiation and freedom of research in some cases."[4] Likewise, the Society for the Study of Social Problems endorsed the spirit of the February policy, but it doubted that local review committees could competently and fairly evaluate the ethics of proposed research. Like the participants at the NIH conference, the Society's members feared that committees would be too sensitive to "political and personal considerations" and insensitive to "important differences in the problems, the data, and the methods of the different disciplines." It called on Surgeon General Stewart to consider alternatives to local IRBs, such as national review panels composed of experts.[5]

The Public Health Service mostly ignored these arguments. Regardless of their contacts with sociologists, PHS officials seem primarily to have had psychology in mind when they spread the prior-review requirement to nonbiomedical research.[6] Interviewed in 1971, Gordon—the psychologist working on the policy—mentioned his concern about both personality tests and deceptive experiments in social psychology, but not about research in anthropology or sociology. And he mentioned both positive and negative feedback from psychologists and psychiatrists, but no contact with anthropologists, sociologists, or other social scientists.[7] In September 1966 Gordon did hint that the PHS might "try to identify the kinds of investigations involving human subjects that pose no risks to their rights and welfare," which might include "anthropological or sociological field studies in which data are collected by observation or interview and which include no procedures to which the subjects, or their guardians or parents, or the authorities responsible for the subjects may object."[8] For such cases, the policy would not apply. But no such exemption made it into the clarification issued at the end of the year.

Instead, on 12 December 1966, the PHS announced explicitly that "all investigations that involve human subjects, including investigations in the behavioral and social sciences" would have to undergo the same vetting as medical experiments. The announcement did claim that only the most risky projects would require "thorough scrutiny." In contrast,

> there is a large range of social and behavioral research in which no personal risk to the subject is involved. In these circumstances, regardless of whether the investigation is classified as behavioral, social, medical, or other, the issues of concern are the fully voluntary nature of the participation of the subject, the maintenance of confidentiality of information obtained from the subject, and the protection of the subject from misuse of the findings. For example, a major class of procedures in the social and behavioral sciences does no more than observe or elicit information about the subject's status, by means of administration of tests, inventories, questionnaires, or surveys of personality or background. In such instances, the ethical considerations of voluntary participation, confidentiality, and propriety in use of the findings are the most generally relevant ones. However, such procedures may in many instances not require the fully informed consent of the subject or even his knowledgeable participation.[9]

Surgeon General Stewart later claimed that this statement addressed the sociologists' concerns, but he offered no evidence that any sociologist agreed. He also pledged that "should we learn that a grantee is stopping research unlikely to injure human subjects, we would express to the grantee our concerns and clarify the intent of the relevant policy."[10]

Even as this policy was being drafted in late 1966, Congressman Henry Reuss (who had endorsed Gallagher's letter of 1965) commissioned anthropologist Harold Orlans to investigate the use of social research in federal domestic programs. As part of a lengthy questionnaire, Orlans asked respondents' opinions of the Public Health Service's prior-review requirement of February 1966, as well as a July supplement in which the PHS had required a promise of informed consent for the administration of personality tests, inventories, or questionnaires. Orlans's questionnaires went to 146 leading social scientists at universities and independent research organizations. Fifty-three responded, of which twenty-eight expressed relatively clear opinions about the PHS's actions. Of those, five psychologists, a psychiatrist, and two anthropologists were reasonably supportive of the policy.

In contrast, twenty respondents—in a variety of disciplines—were opposed, some of them quite bitterly. Sociologist Alvin Gouldner raged, "The Surgeon General's ruling . . . tends to increase surveillance of research by persons who are not fully knowledgeable about this research. I deplore the growth of the bureaucracies which are surrounding social research and feel that they will smother the integrity and the originality of this research."[11] Political scientist Alfred de Grazia foresaw "a depressing tangle in which neither individual right nor free research is helped."[12] And Rensis Likert, director of the University of Michigan's Institute for Social Research, argued that "ethical issues will not be resolved effectively by direct legislation nor by administrative edict."[13] He and other respondents called for continued ethical discussions within professional organizations and better education for researchers to solve what they saw as a relatively minor problem of occasional ethical lapses.

Orlans disagreed. Having solicited the opinions of leading scholars, Orlans dismissed those opinions as naïve, arguing that social scientists had proven themselves incapable—individually and collectively—of self-regulation. He then argued that the only alternatives were ethical review by federal employees or ethical review by a peer committee, and that the latter was less intrusive and therefore preferable. He even claimed that "our typical respondent . . . is sympathetic to the Surgeon General's approach."[14] In fact, he had not found a single economist, historian, political scientist, or sociologist who held such sympathies, and not even all the psychologists he contacted were on board.

A final opinion—this one warmer to the idea of institutional review—came from the 1967 report of the presidential Panel on Privacy in Behavioral Research, chaired by psychologist Kenneth Clark. The panel had been created in January 1966, in part as a response to Gallagher's hearings. It took as its brief the study of privacy issues in "economics, political science, anthropology, sociology, and psychology," though the twelve-member panel included no political scientists or sociologists. It did contain, along with Clark, another psychologist and a psychiatrist, reflecting the dominance of those fields. It also included anthropologist Frederick P. Thieme, one of the two anthropologists to endorse institutional review in response to Orlans's query.

After about a year of work, the panel reported that "most scientists who conduct research in privacy-sensitive areas are aware of the ethical implications of their experimental designs and arrange to secure the consent of subjects and to protect the confidentiality of the data obtained from them." Still, the panelists found a few exceptions, and they therefore recommended a form of institutional review along the lines of the NIH's system. They called for a scientist's research plan to be reviewed by other scientists "drawn in part from disciplines other than the behavioral sciences, [who] can present views that are colored neither by self-interest nor by the blind spots that may characterize the specific discipline of the investigator."

At the same time, the panel warned against a crude application of medical ethics. While endorsing the idea of informed consent, it suggested that "some modification of the traditional concept of informed consent is needed" to allow for deceptive experiments and for unobtrusive observation. And while promoting the idea of privacy, it distinguished that from a universal goal of confidentiality. "The essential element in privacy and self-determination is the privilege of making one's own decision as to the extent to which one will reveal thoughts, feelings, and actions," the panel explained. "When a person consents freely and fully to share himself with others—with a scientist, an employer, or a credit investigator—there is no invasion of privacy, regardless of the quality or nature of the information revealed."

The panel also warned that "because of its relative inflexibility, legislation cannot meet the challenge of the subtle and sensitive conflict of values under consideration, nor can it aid in the wise decision making by individuals which is required to assure optimum protection of subjects, together with the fullest effectiveness of research." Instead, it recommended that "the methods used for institutional review be determined by the institutions themselves. The greatest possible flexibility of methods should be encouraged in order to build effective support for the principle of institutional responsibility within universities or other organizations. Institutions differ in their internal structures and operating procedures, and no single rigid formula will work for all."[15]

In each of these contexts—the NIH conference, the professional organizations' resolutions, the responses to Orlans's queries, and the privacy panel recommendations—social scientists had expressed mild or severe reservations about the Public Health Service policy. Those reservations may have shaped the surgeon general's December 1966 acknowledgement that the risk levels and ethical considerations of social science were distinct from those of medical research. But the social scientists were unable to dissuade the PHS from imposing the requirement of institutional review on their work. Instead, they got only promises that at some future date, the PHS might try to distinguish between those types of research for which prior review was appropriate and those for which it was not, and that it would "express . . . concerns" if an institution unnecessarily blocked research. The PHS kept neither promise.

Nor were social scientists able to confine review to PHS projects. In 1968, DHEW secretary Wilbur Cohen imposed a reorganization, designed in part to bring the PHS under closer control by the department.[16] But the shuffle also put longtime PHS grants official Ernest Allen into the more senior post of deputy assistant secretary for grants administration for the entire DHEW. Not long afterwards, Allen persuaded the department to seek a "uniform policy," meaning, in effect, to impose the PHS's guidelines throughout DHEW.[17] Since this would mean extending the rules to nonhealth agencies, such as the Office of Education and the Social Security Administration, officials "expressed varying degrees of concern with its impact on nonmedical programs."[18] They found a draft document "strongly medical, and rather casual in its treatment of social and behavioral studies." Yet the drafting committee consulted a number of medical researchers, the American Association of Medical Colleges, and the American Psychological Association, while ignoring social science organizations. Despite this lack of input, on 15 April 1971, the institutional review policy was applied to all department-sponsored research.[19] Not long afterwards, when a review-committee member brought up the question of "behavioral and political science research," department representatives insisted that "questionnaire procedures are definitely subject to the Department's policy."[20]

The department did make some efforts to adapt the policy to nonmedical research. In the summer of 1971, it sponsored a series of meetings in which more nonmedical scientists participated.[21] Their input fed into the December 1971 interpretation of the April policy, entitled the *Institutional Guide to DHEW Policy on Protection of Human Subjects*, but known (because of its cover) as the Yellow Book.[22] The Yellow Book was careful to avoid, when possible, strictly medical language; for example, the imagined human subject was now a "patient, student, or client," rather than just a patient. It cited concerns about "discomfort, harassment [and] invasion of privacy" as possible consequences of noninvasive research. Yet it still imagined that most research under the policy was designed to improve the "established and accepted methods" of some kind of therapy or training, which is not the case in much social research. The final pages of the document cited eleven codes of ethics or statements of principles, covering medicine, mental health, psychology, and social work. The new codes worked out by anthropology, oral history, political science, and sociology were ignored.[23]

Thus, as a result of health officials' concerns about intrusive psychological research, the Public Health Service established IRB review as a requirement for a much broader class of nonbiomedical research. Then, thanks to the PHS's position within DHEW, that requirement spread to the entire department. All of this was done over the objections of leading social scientists and scholarly organizations. And social scientists would again be frozen out when research ethics reached the level of congressional concern.

TUSKEGEE AND THE NATIONAL RESEARCH ACT

In July 1972 reporter Jean Heller broke the news of the Tuskegee Syphilis Study, in which the Public Health Service had observed the effects of syphilis on 399 African American men for forty years without offering them treatment.[24] "Even after penicillin became common and while its use probably could have helped or saved a number of the experiment subjects," wrote Heller, "the drug was denied them." DHEW officials declined to defend the project, while members of Congress expressed outrage, with Senator William Proxmire calling the study "a moral and ethical nightmare."[25] The next year, Congress debated several bills to rein in medical research.

Whatever their proposals, the various sponsors saw the problem as one of medical research, not research in general. Senator Hubert Humphrey proposed a national board to "review all planned medical experiments that involve human beings which are funded in whole or in part with Federal funds."[26] Senator Jacob Javits told the Senate he was concerned about "psychosurgery, organ transplants, genetic manipulations, sterilization procedures for criminal behavior, brainwashing, mind control and mind expanding techniques, and, yes, even the very concept of birth and death itself."[27] Senate hearings—before the Subcommittee on Health—in February and March of 1973 likewise emphasized abuses of medicine and therapy. One concern was the delivery of health services, which had failed so miserably in the Tuskegee study, as well as cases where government funds had supported sterilizations of dubious propriety. The Senate's health subcommittee also addressed "a wide variety of abuses in the field of human experimentation."[28] These abuses included the introduction of new drugs and medical devices, as well as surgical procedures. In all of this, the Senate focused on strictly medical issues.

The final bill, however, called for the further study and regulation of "biomedical and behavioral" research. How and why the term behavioral crept in, and what it meant, remain mysterious. As ethicist Robert Veatch noted in testimony before a House of Representatives committee, behavioral research could mean something as narrow as "research in behaviorist psychology" or as broad as "any research designed to study human behavior including all social scientific investigation." While he preferred the latter, he argued that the most important thing was to define the term. "To leave such ambiguity would be a tragedy," he warned.[29] But Congress ignored the warning; neither the bill nor its accompanying report defined the term.

Observers at the time disagreed over which meaning was appropriate. Writing in late 1974, Emanuel Kay argued for the broader construction. He noted the 1973 Senate testimony of Jay Katz, a physician on the Yale law faculty and the editor of a major anthology of materials on human experimentation, which included attention to social research. In his testimony, Katz had flagged four issues he did not think were adequately addressed by current DHEW polices: research with prisoners, research with children, the jurisdiction of IRBs over medical therapy, and, finally, "research in other disciplines," including "psychology, sociology, law, anthropology, and other disciplines." Katz noted that "deceptive experiments and secret observation studies" offended his sense of informed consent, self-determination, and privacy.[30] Kay maintained that by praising Katz, Edward Kennedy and the other senators had embraced his ideas about the necessary scope of the regulation.[31]

Kay's account, though, poses problems. First, Katz did not use the term behavioral research, calling instead for the regulation of "research in other disciplines." Clearly, the senators had gotten their language from some source other than Katz. Second, while Senator Kennedy certainly thought highly of Katz, neither he nor other legislators picked up Katz's explicit interest in the social sciences. In presenting his bill to the House of Representatives, Kennedy praised Katz as "perhaps the world's leading authority on human experimentation," but when he came to list the "serious medical and ethical abuses" which had spurred his concerns, he stuck to biomedical incidents: the Tuskegee study, the sterilization of two teenage girls without their parents' consent, the widespread use of experimental drugs and medical devices, "and the performance of psychosurgery for the purposes of behavior modification without proper peer review and without an experimental protocol."[32]

This last statement about behavior modification supports the alternative interpretation of "behavioral" offered by Sharland Trotter, who covered the hearings for the American Psychological Association's newsletter. Rather than guessing, Trotter simply asked Kennedy staffer Larry Horowitz, a medical doctor himself, about the language. Horowitz told her that Kennedy had added "and behavioral" in response to testimony "about the research-disguised-as-treatment—from psychosurgery to token economies—that goes on in such closed institutions as prisons and mental hospitals." Indeed, the Senate report—which never mentions anthropology or sociology—repeatedly cites concerns about "behavioral control," citing as an example B. F. Skinner's "research into the modification of behavior by the use of positive and negative rewards and conditioning."[33] In November

1974, the Senate Committee on the Judiciary published its own lengthy report on behavior modification. It too suggested that Congress's real concern was with projects in "behavioral research, where the researcher may have virtually complete control over the well-being of the individual subject."[34] While a matter of serious concern, this work was far removed from the research conducted by most psychologists, much less social scientists. As Trotter noted, "on the basis of testimony . . . citing abuses primarily in *clinical* research, psychology as whole has found itself covered by the proposed legislation."[35]

While all this was going on, the publicity over Tuskegee and Congress's concern prompted DHEW officials to think about replacing their departmental guidelines with formal regulations. Although they believed that the 1971 Yellow Book provided all the protection necessary, they realized that members of Congress and the general public would want more than a pamphlet. In August 1972, just weeks after the Tuskegee disclosures, they began circulating memos about the need for the department to publish formal regulations in the *Federal Register* if it did not want to have them imposed directly by Congress.[36]

Within the NIH, the task of monitoring Congress fell to Charles R. McCarthy, a former priest with a double doctorate in philosophy and politics. McCarthy had taught both subjects until 1972, when he joined the legislative staff of the National Institutes of Health. After attending some of the Senate hearings, he warned that some kind of legislation was coming. If DHEW did not come up with something impressive, Congress might establish a separate agency—outside the department—to control human experimentation.[37]

As they discussed how to mollify Congress, DHEW officials never made clear whether they were debating the regulation of just medical research, or medical and behavioral research, or medical, behavioral, *and* social research. For the most part, their discussions assumed that the focus of any legislation and regulations would be biomedical research. This medical focus was indicated by both the recipients—medical officials within DHEW—and the title of the September 1972 memo calling for new regulations: "Biomedical Research and the Need for a Public Policy."[38] Likewise, the group established in January 1973 to study the problem was named the Study Group for Review of Policies on Protection of Human Subjects in Biomedical Research, with membership drawn from health agencies within the department.[39] At other times, psychiatry and psychology crept in.[40]

When they did mention social science, NIH officials worried about overregulation. In September 1973, an NIH memo objected to one bill on the grounds that its insistence on the subjects' informed consent "places unacceptable constraints on behavioral and social science research."[41] Donald Chalkley, the head of the NIH's Institutional Relations Branch, complained that "by imposing on all behavioral and biomedical research and service, the rigors of a system intended to deal with the unique problems of high-risk medical experimentation, [the bill] would unduly hamper low-risk research in psychology, sociology, and education, and unnecessarily dilute the attention given to potentially serious issues in clinical investigation."[42]

The confusion within DHEW is best illustrated by an October 1973 memo draft by Chalkley: "There is general agreement that while much of the early stages of medical, psychological and sociological research can be carried out using animals, man remains the subject of necessity if new diagnostic, therapeutic, preventative or rehabilitative techniques are to be shown effective and safe for human application." Beyond the absurdity of imagining that sociologists test their questionnaires on lab rats, the memo assumes that the goal of research must be therapy, far less common in social science than in medical research.[43]

This indecision about any policy's scope scuttled an early effort to develop a policy applicable to all federal agencies. In late 1973, representatives of several federal agencies had "approached a broad consensus on a Federal policy for the protection of human subjects." But the chief of the Behavioral Sciences Branch of the NIH's Center for Population Research complained that "certain provisions of this document, if applied to behavioral sciences research, might be self-defeating without clear justification."[44] Likewise, the assistant secretary for planning and evaluation objected that "the policies and rules devised to protect human subjects, as well as to advance science, appear to be inadequate in behavioral science investigations, especially with respect to large-scale studies of non-institutionalized normal subjects."[45] The effort died.[46]

DHEW announced its proposed regulations on 9 October 1973. At the core of the proposal was a reiteration of the IRB requirement that had been in place since 1971: "no activity involving any human subjects at risk supported by a DHEW grant or contract shall be undertaken unless the organization has reviewed and approved such activity and submitted to DHEW a certification of such review and approval." The review would have to determine "that the rights and welfare of the subjects involved are adequately protected, that the risks to an individual are outweighed by the potential benefits to him or by the importance of the knowledge to be gained, and that informed consent is to be obtained by methods that are adequate and appropriate."[47]

The proposed regulations did not define human subjects, but they did suggest that the policy applied to "subjects at risk," defined as "any individual who may be exposed to the possibility of injury, including physical, psychological, or social injury, as a consequence of participation as a subject in any research, development, or related activity which departs from the application of those established and accepted methods necessary to meet his needs." This definition made no distinction among biomedical, behavioral, and social research. Two hundred comments arrived, some suggesting "limiting the policy to physical risks only [or the] differentiation of biomedical risks from behavioral risks."[48] Within DHEW, officials voiced similar concerns. In May 1974, the Office of the Assistant Secretary for Planning and Evaluation suggested eliminating social science research from the regulations.[49] A memo prepared just after the regulations were finalized noted that some agencies within DHEW were fretting "about the extent to which these regulations would inhibit social science research."[50]

The department had little chance to respond to such critiques, since it feared Congress would strip DHEW of its powers entirely, entrusting human subjects protections to a new, independent federal agency.[51] As McCarthy later conceded, "because of the time pressure imposed by Senator Kennedy, the customary DHEW clearance points for the issuance of regulations were either bypassed or given extremely brief deadlines. The result was a set of flawed regulations."[52] DHEW official Richard Tropp complained, "Under the Senate legislative deadline pressure, the secretary published the regulation even though interagency differences within HEW remained tense and unresolved. It was understood within the department—and alluded to in the regulation's preamble—that further negotiation would follow, to produce a consensus regulation crafted so as to be appropriate for all types of research. The urgency of the issue to the secretary's office waned, however, and that has never happened."[53]

The final regulations were promulgated on 30 May 1974, becoming Part 46 of Title 45 of the Code of Federal Regulations, abbreviated as 45 CFR 46. They succeeded in their main mission: persuading Congress to let the department keep control of its own research. Just weeks after the publication of the new regulations, Congress passed the National Research Act, which gave the secretary of Health, Education, and Welfare the authority to establish regulations for IRBs, the very regulations which DHEW had just promulgated.[54]

The two initiatives—one legislative, the other executive—were largely complementary. But the law diverged from the DHEW proposals in two ways. First, the DHEW regulations applied only to research supported by grants and contracts from that department.[55] In contrast, the federal statute required institutions receiving PHS grants to have IRBs to review human subjects research without specifying whether they must review research *not* funded by the PHS.[56] In subsequent years, the department would claim that this provision required IRB review of all human subjects research at federally supported institutions (i.e., every research university in the country), even projects that received no federal funds. As the NIH's Chalkley put it, "this is another aspect of the 'foot in the door' policy by which receipt of Federal funds is made contingent on broad compliance with policy or regulations."[57] Such policies were indeed becoming common. Not long afterwards, Yale president Kingman Brewster lamented, similar policies used

> the "now that I have bought the button, I have a right to design the coat" approach. Thus if we are to receive support for physics, let's say, we must conform to federal policies in the admission of women to the art school, in the provision of women's athletic facilities, and in the recruitment of women and minorities, not just in the federally supported field, but throughout the university. Even in the name of a good cause such as "affirmative action," this is constitutionally objectionable.[58]

The foot in the door worked. Each institution receiving DHEW research funds was required to submit an assurance pledging to abide by the department's IRB policy. By late 1974, more than 90 percent of the assurances filed with DHEW promised to review all research, regardless of funding.[59]

The second discrepancy concerned the types of research subject to review. The law passed by Congress limited its scope to "biomedical and behavioral research." DHEW's regulations, by contrast, applied to all department-funded "research, development, and related activities in which human subjects are involved."[60] Rather than limiting affected research to the biomedical and behavioral categories, the announcement of the regulations explicitly recognized what might be a third category, promising that "policies are also under consideration which will be particularly concerned . . . with the subject of social science research."[61]

Careful observers saw a mismatch. Ronald Lamont-Havers, who had chaired the DHEW Study Group on the Protection of Human Subjects, had earlier distinguished among four categories of human subjects research: biomedical, behavioral and psychological, societal, and educational and training.[62] Attuned to the differences among these areas, he noticed Congress's limit and assumed that it applied to the regulations as well. In September 1974 he noted that the policy "is presently applicable only to biomedical and behavioral research" and argued that "the inclusion of 'social' . . . and 'social sciences' " in two sections of a draft departmental manual "is in error since the policy should deal only with biomedical and behavioral research."[63] An NIH legal advisor was less certain. He replied that "it would seem that the Chapter's references should at least be internally consistent, and perhaps consistent with Part 46 [of the regulations] as well," thus hinting at a discrepancy between those regulations and the statute.[64] But this discussion seems not to have led to any reexamination of the regulations themselves. Technical amendments published in March 1975 failed to resolve the issue.[65]

Embarrassed by the Tuskegee scandal and fearful of heightened congressional oversight or the establishment of a rival agency, DHEW had rushed the regulations into print. It had ignored warnings from within the department and from outside commentators that the regulations might prove inappropriate for the social sciences, and it had diverged from the concerns of Congress and the law itself. It had written regulations that were highly ambiguous in their treatment of social research, and the effects of this haste would soon be felt at universities across the country.

EARLY IRB REVIEW OF THE SOCIAL SCIENCES

Because the Public Health Service sponsored relatively little social research, its 1966 policy seems not to have produced routine IRB review of such research, much less any controversy about IRBs. A 1968 Public Health Service memorandum noted more confusion than anything else: "We have had questions from medical school committees questioning whether there was any need to review projects in psychology, from psychologically oriented committees, questioning the need to review anthropological studies, and from everyone else questioning the need to review demonstration projects."[66] The response was that such projects did need review, but that the policy needed clarification "to avoid the necessity of obtaining assurances from the YMCA, the PTA, and the Traveler's Aid Society."[67]

Laud Humphreys's research gives a hint of the confusion of the period. As the recipient of two fellowships from the National Institute of Mental Health (a Public Health Service component), he was subject to the PHS's 1966 directive requiring prior review by a committee. But no one seems to have noticed this until after he had defended his dissertation, at which point the university chancellor chastised him and raised the specter that his degree might be revoked. Yet all it took to end the controversy was a statement by the department secretary that she was to blame for failing to submit the appropriate forms.[68] This suggests a certain casualness of enforcement due to the newness of the rules.

Into the early 1970s, the requirement of IRB review of social and behavioral research raised relatively little controversy. Interviewed in February 1971, Mark Conner of the NIH's Office of Institutional Relations reported scattered opposition by social psychologists, sociologists, and anthropologists, but he was unaware of official opposition by a scholarly organization.[69] And in March 1972, the American Sociological Association's executive director commented on the 1971 Yellow Book: "We haven't had much flak about it so far. But we do have some concern that this could become a political football, shoving aside scholarly concerns."[70] As late as 1976, a sociologist checking with colleagues around the country found that many did not know IRB rules might apply to their work.[71] At Princeton, political scientists simply did not submit proposals to their IRB.[72]

At other universities, however, concerns grew as the DHEW regulations went into effect and as the department stepped up its enforcement. In late 1974, the NIH renamed its Institutional Relations Branch as the Office for Protection from Research Risks (OPRR), expanding its staff and upgrading its place on the NIH organizational chart.[73] Though as late as 1977 the office had only fifteen people on staff (none of them social scientists), its promotion signaled increased clout.[74] And the OPRR was willing to use that clout to impose its ideas about internal review. For example, at some point in the 1970s, Ohio State University established an IRB with a subcommittee governing research in each department, presumably out of respect for disciplinary differences. The OPRR quashed the idea, insisting that all behavioral research be reviewed by a single board.[75]

Such interventions exacerbated tensions within universities. Perhaps the earliest fight over IRB review of social science took place at the University of California, Berkeley, a leading research university with a prominent sociology department. Berkeley had formed its first human subjects committee in 1966, in response to Public Health Service requirements. In 1970 the university president applied the PHS's standards to all human subjects research, regardless of funding, and in March 1972 the DHEW requirements became the standard.[76] Even undergraduate projects would now be monitored by Berkeley's IRB, known as the Committee for the Protection of Human Subjects.[77]

Real controversy began in November 1972, when the university administration announced that research would be vetted not only for risks to individuals but to groups and institutions:

> Even when research does not impinge directly on it, a group may be derogated or its reputation injured. Likewise, an institution, such as a church, a university, or a prison, must be guarded against derogation, for many people may be affiliated with, or employed by, the institution, and pejorative information about it would injure their reputations and self-esteem.

The author of the policy, IRB chairman and law professor Bernard Diamond, later explained that he had been thinking of the work of Arthur Jensen and William

Shockley, who had suggested that African Americans' genetic inheritance made them, on average, less intelligent than whites. But University of Chicago sociologist Edward Shils warned that the policy would also prevent researchers from exposing lazy policemen, corrupt politicians, and monopolistic businessmen, among others. Berkeley researchers agreed, noting that a policy directed at "preventing, in practice, scientific investigation of certain unpopular topics" could be turned on anyone. They formed an Academic Freedom Committee in protest.[78]

In between the two sides was anthropologist Herbert Phillips, a member of Diamond's IRB. Phillips, a specialist in Thailand, had been dismayed by the American Anthropological Association's handling of the debates over clandestine research in Southeast Asia during the Vietnam War. A voluntary membership organization, he believed, had no business trying to police individual accusations of misconduct. Rather, he thought that interdisciplinary human subjects committees would be best able to ensure ethical research.[79] He also supported the concept of informed consent, arguing that "a researcher's obligations to his human subjects are as great as his obligations to scholarship or science," and that subjects had the right to know a researcher's aims and hypotheses. Thus, if another Jensen came along seeking to test more African American children, their parents should be informed about what the study hoped to prove.[80] But Phillips was equally insistent in rejecting Diamond's concept of social risk, arguing for "the protection of human subjects, not of social groups."[81] After a long and vigorous debate, this compromise held, and Phillips took over as the chair of the Committee for the Protection of Human Subjects.[82]

Phillips turned his committee into a model of discretion and accountability. He unilaterally approved projects he felt presented no real risks, or else hurried them through the committee. When the Berkeley political science department—which was doubtless conducting surveys—failed to submit any protocols for review, Phillips ignored it, on the grounds that no one would be hurt by a political survey. For protocols that were submitted, Phillips waived the requirement for written informed consent when the work posed little risk or when the subjects, such as public officials, could be expected to protect themselves.[83] Instead, he tried to present the full committee only with cases that were truly challenging or perilous, and then he held his committee to high standards of transparency.[84] All deliberations were tape recorded, and the researcher was given the opportunity to listen to the recording as a protection against judgments being made based on "bias, whimsy, or incompetence." Rather than reinventing the wheel with each new proposal, the committee cataloged each application, allowing it to develop an internal case law based on experience.[85] At times the committee offered suggestions that promised to improve protection without compromising research. For example, in a study of illegal drug use by teenagers, the committee advised the researcher to protect the code key—which could link names of respondents to data about them—from subpoena by getting an exemption from the attorney general or by keeping the key in a foreign country.[86]

As time went on, however, Phillips became frustrated by the DHEW requirements. He found that 85 to 90 percent of the projects he reviewed needed no modification, but still required an "immense amount of meaningless paper shuffling." He doubted that prior review of protocols would be particularly helpful, since methods inevitably change in the course of a project. Most importantly, he felt the whole scheme put too much responsibility on the committee and not enough on the researcher. As Phillips explained, if the committee was going to trust researchers to say what they were doing with the people they studied, it might as well trust them to determine whether those same people were at risk.[87]

By early 1974, enough researchers were fed up with the policy that they began thinking about other options, and Berkeley law professor Paul Mishkin offered his university an alternative.[88] Rather than placing every proposal before the committee, researchers would be given a handbook, based on 2,500 actual cases already decided by that committee. If, having read the handbook, the researcher concluded that he was not placing anyone at risk, he would then file an affidavit to that effect and proceed with his work. The human subjects committee would then have more time to deal with the 10 or 15 percent of projects that raised serious concern, including all projects involving the mentally infirm, prisoners, and other vulnerable groups. The alternative plan was approved by the university's administration, but it had not yet taken effect in July 1974 when Congress passed the National Research Act. The following spring, the OPRR interpreted that law to require each IRB to determine which projects were risky. It also asserted that review was required for all human subjects projects, whether or not they were funded by DHEW. In short, the Berkeley alternative was stillborn. The only concession the OPRR made was to exempt classroom projects by students.[89]

By 1975, Phillips had become sufficiently outspoken that scholars at other universities began seeking his advice when confronted with ethical charges. He was particularly outraged by DHEW's treatment of two psychologists denied funding by the National Institute of Mental Health. Though both psychologists' proposals had already been cleared by the researchers' own IRBs, the NIMH review panels (whose main task was to judge what projects were most worthy of funding) flagged both proposals for possible ethical violations. In both cases the review panels rejected the proposals in part because the research findings might damage public perceptions of the groups—African Americans and immigrants to England—to which the subjects belonged. In the first case, the director of the NIMH's Division of Extramural Research Programs assured the researcher that the committee's evocation of group harms was a mistake, since DHEW regulations did not cover risks to whole classes of people.[90] But in the second case, Donald Chalkley, the director of the OPRR, stated that both NIH peer review panels and local IRBs were free to consider the social policy implications of the potential findings of research.[91]

Phillips considered these two cases indicative of broader problems with DHEW's approach. First, it had allowed its committee to use a definition of social risk that Phillips found "conceptually obtuse, morally patronizing, and frighteningly anti-intellectual."[92] And second, they showed that DHEW lacked a clear policy, or even a means by which to learn from its mistakes. He called on DHEW to gather information from IRBs around the country to develop a national consensus on research ethics.

Phillips remained a supporter of ethics committees, in principle. But when, in August 1975, he returned to Thailand for more fieldwork, he did not submit a protocol to his old committee. He felt confident that his projects, which involved prominent Thai artists and intellectuals, posed no ethical problems and that he knew Thailand far better than the committee, which included some "idiots." Committee review, he believed, might still be appropriate for graduate students preparing for their first trip into the field, but it would be just waste time for an expert like him.[93]

A second storm took place at the University of Colorado at Boulder. In 1972, that university submitted a general assurance to DHEW—pledging to abide by its policies—and the university circulated memos to the faculty and students telling them that all projects, regardless of funding, should be submitted to the Human Research Committee (HRC) for review. Researchers largely ignored such pleas, neither informing the committee about nonfunded work nor attending its policy meetings. A few departments took up the committee's offer to form departmental subcommittees, which would have the power to review the most common types of research done in those departments. In 1975, DHEW requested a new assurance and then warned the university that "Public Law 93-348 requires that the IRB (Committee) once it is established, review biomedical and behavioral research conducted at or sponsored by (the institution) to protect the rights of human subjects of such research; there is no provision for exceptions."[94] Since the regulations implementing that law stated their applicability only to DHEW grants and contracts, this was a shaky interpretation, but it was consistent with other DHEW actions.[95]

The Colorado administration responded with a new assurance, dated February 1976, pledging that its HRC would "review all activities involving human subjects in all fields of University activity—medical, behavioral, and social (not including projects which will normally come under the purview of the University Medical Center Committee)." Departmental subcommittees might still collect information on some projects, but only the IRB would be allowed to determine risk.[96] In April the HRC pronounced that "every experiment involving human subjects (including questionnaires, surveys, unobtrusive observation experiments, and the use of public information) must be reviewed," emphasizing that "*all* studies, funded or nonfunded, must be reviewed, whether risk is involved or not."[97] Failure to comply could lead to the denial of university facilities, of legal aid, and, most importantly, of graduate degrees.

This announcement awoke the somnolent faculty and sparked a campuswide debate. Sociologist Edward Rose pledged that his doctoral students' dissertation work was harmless—the students would keep away from confidential information and informants' "statements damaging to themselves and to others."[98] And, he pledged, each student's doctoral committee would provide ethical oversight. He then objected to IRB oversight as "censorship" and an intrusion on the "privileged relation between a teacher and his students." As a tenured professor on the eve of retirement, he himself had little to fear from the human subjects committee, and he had not submitted any projects for review.[99] But he knew his students would be far more vulnerable, and he threatened to tell them to seek new advisers rather than submit to IRB review.[100]

The controversy grew bitter. After attending a meeting of the sociology department, IRB chair William Hodges, a psychologist, accused the sociology faculty of indifference to the rights and welfare of their human subjects.[101] In reply, sociologist Howard Higman accused Hodges of harboring a premodern belief in censorship.[102] Rose complained to the university's Committee on Privilege and Tenure, to the chancellor of the university, to the attorney general of Colorado, and to the secretary of Health, Education, and Welfare.[103] The last complaint provoked DHEW to respond in very insistent terms that yes, even non-federally funded social research was subject to review.[104] As at Berkeley, angry faculty formed an academic freedom committee to protest the actions of the human subjects committee.[105]

In early 1977, tempers cooled. Hodges apologized for stating, in the April 1976 memo, that research based on public information would require review; he had meant to require review only for private records, such as those generated about individuals by schools and hospitals.[106] He signaled his committee's willingness to compromise with researchers on such issues as deception and informed consent. Higman was pleased with the concessions and reported that the two committees "seem to be approaching a satisfactory accord."[107] But Hodges stood his ground on the question of his committee's authority over social science work, and he recommended that the dean not grant any spring 1977 graduate degrees to "any student who knowingly and deliberately refuses to submit their thesis study involving human subjects for the review of our Committee."[108] The dean of the graduate school agreed, and in April 1977 he threatened to withhold the doctoral degree from Rose's student, Carole Anderson, unless she received IRB approval.[109]

Eventually, however, the IRB found a way to grant Anderson her degree without ceding any authority and without requiring any action on her part. Two members of the IRB read her dissertation and found that it did not put subjects at risk, and the IRB chair confirmed this finding by interviewing one of Anderson's subjects. Satisfied, the committee withdrew its bar on her graduation.[110] But it warned that it would read no more dissertations. Henceforth, students would simply have to comply. The debates continued into the summer and beyond.[111]

Though in the end no one's career was sacrificed, the University of Colorado storm suggested that the peace of the early 1970s was due less to social scientists' acceptance of human subjects policies than to IRBs' failure to enforce them. Anderson completed her degree, and neither at Berkeley nor at Colorado had IRBs seriously interfered with social research. As Colorado's Hodges took pains to point out, his IRB sought not to "prevent or prohibit" research, only to review it. The academic freedom committees at both universities complained primarily about potential abuses by IRBs, not actual events; perhaps they overreacted. That said, the two IRBs had given scholars good reasons for alarm. Berkeley's warnings of "social risk" and Colorado's initial insistence on reviewing work with public records suggested that the committees planned to go far beyond the duties assigned by federal regulations and enter the realm of outright censorship. Though the two committees retreated from these positions, they had trouble reestablishing civility and trust.

Ironically, in a third controversy over the scope of the 1974 regulations, DHEW got a taste of its own medical ethics. In the fall of 1975, the state of Georgia, looking for ways to cut down on Medicaid costs, wanted to try imposing copayments for doctor visits, hospital stays, and other medical services. In November, two Medicaid recipients sued to block the program on the grounds that the secretary of Health, Education, and Welfare, David Mathews, had not followed the approval procedures set by the Social Security Act. When that argument failed to win a preliminary injunction, the plaintiffs came up with a new argument: the copayment program was a research project involving human subjects, and therefore could not proceed until approved by an IRB. This put Mathews in a tough spot. Apparently, he did not want his work delayed or prohibited any more than did the scholars at Berkeley and Colorado, so he and his human subjects chief, Donald Chalkley, argued that the program did not require IRB review.

To make this case, Mathews and Chalkley put forth some of the very arguments made earlier by such critics as Phillips and Rose. They argued that the Medicaid recipients were not at risk. True, they conceded, the recipients might lose some money to the copayments, but such loss did not count as harm under the regulations. At the same time, however, Mathews and Chalkley did think that a survey of doctors and patients, which had been planned as part of Georgia's program, did put subjects at risk, since "survey instruments designed to test attitudes may have psychological impact upon the interviewees by, for example, inducing in them a sense of shame, guilt, embarrassment, or other emotional reaction." Judge Charles Moye ruled such arguments nonsensical, writing that "to find that the surveys are within the scope of the regulations because the interviews may have a psychological impact upon the interviewees, but that the actual imposition of co-payments are not within the scope of the regulations defies logic." And in any event, the regulations required that an IRB, not the investigator, determine who was and who was not at risk. He ordered that unless the program received approval by an IRB within forty-five days, it would be permanently blocked.[112]

Mathews responded by seeking to free himself of the rules his department imposed on others. Two weeks after the judge's order, he announced in the *Federal Register* a new interpretation of the regulations, claiming that "types of risk situations against which the regulations were designed to protect are suggested by the areas of concern which were addressed in the legislative hearings" of 1973. These, he noted,

> included the use of FDA-approved drugs for any unapproved purpose; psychosurgery and other techniques for behavior control currently being developed in research centers across the nation; use of experimental intrauterine devises; biomedical research in prison systems and the effect of that research on the prison social structure; the Tuskegee Syphilis Study; the development of special procedures for the use of incompetents or prisoners [as subjects] in biomedical research;

> and experimentation with fetuses, pregnant women, and human *in vitro* fertilization. The regulations were intended, and have been uniformly applied by the Department, to protect human subjects against the types of risks inherent in these types of activities.

Therefore, he argued, projects like the Georgia Medicaid program were not covered.[113]

Whatever the merits of such a policy, Mathews's statement was a poor description of DHEW's previous actions. If DHEW had confined itself to the matters addressed in the 1973 hearings, it would have drafted very different regulations in 1974. It would not have intervened in social science research at Berkeley, Colorado, and elsewhere, nor would Mathews have told Judge Moye that the regulations covered the danger of "psychological impact" from survey research. And within months of Mathews's proclamation, Chalkley was ignoring it, arguing (in correspondence to University of Colorado researchers) that IRBs should review questionnaire and survey research, because "the extraction of information may, if the respondent is identified, involve some invasion of privacy and, even if invasion of privacy is not involved, present some psychological risks if the nature of the questions is such as to constitute a threat to the respondent's system of values and adjustment to society."[114] No mention here of the areas of concern addressed in the 1973 hearings. Chalkley did concede, however, that "the present regulations are overbroad and that an effort must be made to narrow their applicability."[115]

Faced with the reality of IRB jurisdiction, even the secretary of Health, Education, and Welfare and the head of the Office for Protection from Research Risks began to have doubts about the breadth of the regulations. And social scientists, on their part, were beginning to wake up to the threat posed by IRBs. Some scholars, such as Herbert Phillips, who had begun the 1970s in favor of prior review and who had put in long service chairing his IRB, had come to think that many projects should never come before such a board.

For most of the first decade of human subjects oversight, from 1966 to 1976, social scientists had little chance to shape policy. They had dutifully responded to queries from Congress, the Public Health Service, and the Department of Health, Education, and Welfare, only to have their recommendations ignored and, at times, misrepresented. At the congressional hearings concerning abuses in human subjects research, they were not asked to testify about their work. At Berkeley and Colorado, their efforts to deploy the traditional tools of faculty governance were overruled by federal officials. Then, in 1974, IRB skeptics gained a new venue for their complaints: a federal commission charged with developing policy recommendations on human subjects research. Unfortunately for the critics, the members of that commission proved as indifferent to the concerns of social scientists as all the policy makers who had come before them.

CHAPTER THREE

THE NATIONAL COMMISSION

The National Commission for the Protection of Human Subjects of Biomedical and Behavioral Research met from December 1974 through September 1978. In those four years, harried commissioners and their small staff reported on an impressive array of topics: research on vulnerable populations, specific medical techniques, and general ethical questions. The commission's claims about the ethics of human subjects research, and its recommendations, remain the foundation for the regulation of much scientific research in the United States and around the world, including social science research.

Yet accounts are split over the degree to which the National Commission ever intended to regulate the social sciences. In 1979, for example, Robert Levine—a medical professor who had written several papers for the commission—insisted that the "Commission was quite concerned with problems of social and behavioral scientists . . . As it considered each recommendation it was forced to think, 'What would be the implications of this recommendation if it were to be applied to social research?' "[1] That same year Bradford Gray, who served on the commission staff, offered the more tepid assessment that the commission "has not been insensitive to issues brought to its attention . . . by social researchers who have written to the Commission or appeared at its hearings."[2]

In contrast, in a series of interviews in 2004, other participants emphasized their biomedical perspective. Commissioner Dorothy Height explained, "Our task had to deal with medical and behavioral, and I think that we stayed within the context of our assignment. To me we did not go beyond that." Commissioner Karen Lebacqz recalled, "We were a Commission that was supposed to look at both biomedical and behavioral research. But, through almost all of our reports, we used the biomedical model as our primary model." Tom Beauchamp, a commission consultant, concurred: "The truth is, very, very little attention was paid to social science and behavioral research. Very little that we wrote pertained to that."[3]

This chapter and the next use the commission's own records to present an account somewhere between the two interpretations, though much closer to the more recent version stressing the commission's inattention to the social sciences. Those records show that the commission began its work firmly focused on the ethical problems of biomedical research alone. Over time, commission staffers and consultants, Levine and Gray included, did become aware of some of the implications of the commission's work for the social sciences (though not the humanities). But this awareness came too late to affect the basic shape of the commission's work, and it was confined mostly to the staff level, with the commissioners expressing little interest in the issue. Rather than a full consideration of the rights and responsibilities of social scientists, the commission offered only small adjustments to its fundamentally biomedical recommendations.

THE MEDICAL ORIGINS OF THE COMMISSION

The National Research Act established the commission as an expert body scheduled to work for two years. It would be unusually powerful: the designated recipient of its recommendations—the secretary of Health, Education, and Welfare—could not legally ignore them. Rather, the secretary was required to print each report in the *Federal Register*, collect comments, and—within 240 days of its receipt—either implement a recommendation or explain his reasons for judging it inappropriate.[4]

Thc law instructed the commission to study issues concerning research using children, prisoners, the "institutionalized mentally infirm," and living fetuses, as well as psychosurgery. The term "and behavioral" was sprinkled throughout the commission's mandate, so that the Senate's original call for the commission "to identify the basic ethical principles which underlie the conduct of biomedical research involving human subjects" became a Congressional mandate "to identify the basic ethical principles which underlie the conduct of biomedical and behavioral research involving human subjects."[5]

The makeup of the commission reflected the concerns raised in the Senate's post-Tuskegee hearings. The act specified that the commission should be composed of

> individuals distinguished in the fields of medicine, law, ethics, theology, the biological, physical, behavioral and social sciences, philosophy, humanities, health

> administration, government, and public affairs; but five (and not more than five) of the members of the Commission shall be individuals who have been engaged in biomedical or behavioral research involving human subjects.[6]

This definition left doubt about whether the social sciences were distinct from the behavioral sciences. Perhaps social scientists belonged in the same role as theologians, lawyers, and philosophers: included among the professions expected to have valuable opinions, but not among the human subjects researchers whose work would be the object of the commission's recommendations. Or perhaps they belonged with biological and behavioral scientists, who were expected both to contribute guidance and to be subject to that guidance.[7]

The matter could have been cleared up had the secretary of Health, Education, and Welfare appointed a social scientist to the commission—forcing the issue of whether he or she was to occupy one of the researcher slots. (In 1978, a law establishing a later commission avoided this problem by distinguishing slots for biomedical or behavioral research from slots for scholars in the social sciences.)[8] In practice, by specifying at least thirteen broad areas of human expertise to be represented by only eleven commissioners, Congress guaranteed that not all of the desired fields would be included, and social science was one of the areas left out. Of the five researchers on this commission, three (Robert Cooke, Kenneth Ryan, and Donald Seldin) were physicians, and two (Joseph Brady and Eliot Stellar) psychologists. Although a minority on the commission, these researchers would dominate its deliberations, with the three physicians being particularly influential.[9] Ryan, an obstetrician, was elected chairman at the first commission meeting. The remaining six commissioners were three lawyers, two ethicists (Karen Lebacqz and Albert Jonsen), and a social worker. Most or all of the commissioners owed their appointments largely to their expertise in specific issues mentioned by the law—such as psychosurgery, research with children, and research with pregnant women.[10]

Of the eleven commissioners, perhaps only Jonsen had a strong understanding of social science methods and ethics. As an undergraduate, he had been interested enough in anthropology to spend two summers conducting fieldwork on an Indian reservation in Montana, and he later considered pursuing graduate work in anthropology. But Jonsen had gained his seat on the commission for his expertise in biomedical research ethics, and that is where he focused his attention. As he later explained,

> I didn't feel I was battling for [social] science. The problem that I thought was at the very essence of biomedical research did not seem to me to exist in the social

> sciences. And that's the conflict of interest between physician as treating physician and physician as researcher. I just didn't see that as a problem in the social sciences at all.[11]

Though DHEW sought nominations from the American Sociological Association and the American Anthropological Association, in the end no sociologists, anthropologists, or other social scientists were appointed to the commission.[12] (This did not, however, stop the department from claiming that it had found experts in all the categories specified by Congress.)[13] Thus the commission began its work with every reason to think that its main job was to explore the ethics of medical experimentation on human beings, with some attention to the question of behavioral modification. It would be pulled away from those important tasks only with great effort, and even then, not very far.

The early work of the commission reflected the biomedical focus of the Senate hearings and the resulting 1974 National Research Act. After some administrative work, such as selecting a chair and setting a schedule of meetings, the commissioners plunged into the first controversies on their list—research on the fetus and psychosurgery. This haste to wrestle with specific biomedical challenges precluded consideration of just what was meant by behavioral science or behavioral research, terms the commission never defined.[14] However, the commission's early work was consistent with the congressional instruction to delineate the "boundaries between biomedical or behavioral research involving human subjects and the accepted and routine practice of medicine." Since this provision assumed that behavioral research had a boundary with medicine, the law might be taken to imply that it consisted of psychology and psychiatry. At the commission's June 1975 meeting, commissioner Donald Seldin proposed that the question before the commission was "how you intervene in the behavior patterns of people, according to a medical model." He wanted to make sure that the commission did not wander into studies of advertising or other, less direct shapers of behavior.[15]

Early staff papers concurred. One paper identified several psychiatrists and psychologists as behavioral scientists. One of these scientists, psychiatrist Don Gallant, offered a wordy definition:

> Biomedical or behavioral research involving human subjects should be defined as well-designed and critical investigations of therapeutic techniques with unknown efficacy and/or risks or an attempt to find the etiology of a disease having for its aim the discovery of new facts associated with the 'accepted and routine practice of medicine' with the ultimate goal of providing beneficial effects for human subjects.[16]

In an October 1975 paper on research in prisons, the commission staff offered an even narrower view, defining behavioral research as "any program instituted on an experimental basis involving innovative behavior modification or other behavior-related techniques."[17] Both papers conceived of behavioral research as a means to improve behavioral therapy, much as biomedical research had, as its ultimate goal, better medical therapy. Through the end of that year, then, the commission stayed close to the concerns that created it—the medical and psychological abuses that Congress had investigated in 1973.

Yet the scope of the commission's work remained fuzzy. This fuzziness can best be seen in the work of Robert Levine, the physician and professor of medicine at Yale who wrote several key papers for the commission. Levine began by dismissing social research entirely, such as when tasked to explore the National Research Act's call for the commission to consider "the boundaries between biomedical or behavioral research involving human subjects and the accepted and routine practice of medicine."[18] In a November 1974 presentation, Levine reported that in meeting this requirement "we may exclude those types of research which by their very nature have no boundaries with the practice of medicine (including psychiatry). Thus, under this rubric, we need not consider social research."[19]

By December, however, Levine thought that while it did not belong in the boundaries discussion, "social research must be considered in other parts of [the National Research] Act."[20] And by late June 1975, Levine decided to drop the problem in the lap of the commissioners, telling them, "Until I know whether or not the Commissioners want me to cover social science, education and welfare, there is no chance you are getting a draft of anything. I can either cover it or not cover it." The commissioners rejected the idea that new policies in education and welfare needed Levine's attention, but no one objected to commissioner Eliot Stellar's proposal to include the social sciences. Although the commissioners did not take a formal vote, chairman Kenneth Ryan noted "the general Commission feeling that we are talking about biomedical, social, and behavioral research."[21]

When Levine submitted his report two weeks later, he included a section on social science. While acknowledging that most social scientists "have no professional practices other than research," he argued that a criminologist working with a law enforcement agency could be compared to a medical researcher practicing medicine. More broadly, social scientists "may develop knowledge that will be used to 'remedy' the societal dysfunctions they have 'diagnosed.' Thus, in this fashion, their role might be compared to that of the physician with society as the patient-subject. Thus, perhaps in some cases, it is society that we ought to offer the opportunity to give informed consent." Had Levine probed these medical analogies, he might have concluded that a law enforcement officer has a rather different ethical relationship to a criminal than a doctor to a patient. He might also have asked under what circumstances a society can withhold consent to being studied, and if this was anything but a fancy term for censorship. But that was not his purpose. Rather, he contented himself with fitting the social sciences into medical categories—however awkwardly. Levine's uncertainty affected his proposed definition of human subjects research:

> Research (involving humans) is any manipulation, observation, or other study of a human being—or of anything related to that human being that might subsequently result in manipulation of that human being—done with the intent of developing new knowledge and which differs in any way from customary medical (or other professional) practice. Research need not be interactive; e.g., observations of humans through a one-way glass, by tape-recording their conversations with each other, or by examining their records may—but need not necessarily—be classified as research.[22]

The final hedge—need not necessarily—perhaps recognized that in the previous sentence, Levine had defined human subjects research so broadly as to include looking up a number in a telephone book.

Levine paid more attention to social science as such in a second report, submitted in September 1975, about the utility of risk-benefit analysis. Here he expanded his earlier understanding of social research as inherently therapeutic, noting that some social science "is basic research involving human subjects (done with the intent of improving our understanding of the structure and function of various societies, institutions, and so on)." He also noted that the risks and benefits of such work would be hard to define: "Some anthropologists by joining a society may actually change the nature of the society. There may be considerable debate as to whether the changes are detrimental or beneficial." Likewise, he wrote, "it is difficult to assess [the benefit of such work] either quantitatively or from a utilitarian point of view. But how many people who will never go to Samoa have enjoyed M. Mead's books? How many are fascinated by the *National Geographic*?"

Much social science work, then, had unknowable—and surely unquantifiable —benefits. Levine hinted that this unknowability might pose problems for an IRB trying to follow the DHEW requirement that it determine whether "the risks to the subject are so outweighed by the sum of the benefit to the subject and the importance of the knowledge to be gained as to warrant a decision to allow the subject to accept these risks."[23] He suggested instead that in social science research "the IRB should determine that there is an adequate description of risks and benefits (as broadly defined in this paper) to permit the prospective consentor to make a rational decision."[24] In the space of a year, Levine had moved from dismissing social research as outside the purview of the commission, to forcing it into medical categories, to understanding it, at last, as a human activity so distinct from medicine that it might demand its own ethics. Levine later backtracked a bit. In a July 1976 paper he argued against a hard boundary between biomedical and behavioral research without addressing the possibility of a boundary between behavioral and social research, or between social research and such human enterprises as history, journalism, and fiction writing.[25]

The commissioners themselves—none of them social scientists—remained uninterested in these questions, barely mentioning the social sciences throughout 1975 and not much after that.[26] In February 1976, the commission staff defined research as "a class of activities designed to develop or contribute to generalizable knowledge," in contrast to "the practice of medicine or behavioral therapy [that] refers to a class of activities designed solely to enhance the well-being of an individual." Like Levine's papers, this definition responded to the congressional charge that the commission distinguish research from therapy. But this definition did nothing to separate out the types of research that should be subject to IRB review from those activities, such as reading books, that did not require prior review. The commissioners discussed the implications of their definition in strictly medical terms. For example, Donald Seldin, MD, stated that "the underlying purpose of all of this is, in some sense, the relief of pain and disability and the prevention of death, and ultimately one of the most precious instrumentalities to accomplish this is in fact research." He then contrasted the introduction of coronary bypass grafts with modified drug dosages to explain the need to distinguish truly transgressive procedures from minor innovations.[27] However often social science was mentioned, the commissioners seemed unable to keep it in mind.

This biomedical focus posed a challenge to the only social scientist with a major role in the commission's work, the sociologist Bradford Gray. Gray had earned his doctorate in sociology at Yale with a study of review boards at a Yale University hospital, published in 1975 as *Human Subjects in Medical Experimentation*. He found that the boards (chaired by Levine) helped promote ethical research, but that their prior review of research was insufficient to ensure truly informed consent, and he called for boards to interview randomly selected research subjects to find out if they really understood what they had volunteered for.[28] Since this work was one of the few studies of an IRB up to that point, it made Gray a leading expert on the subject and a natural choice for a staff position.[29]

Gray was a sociologist of medicine, not a sociologist of sociology or any kind of expert on the ethics and methods of social science. As the commission stumbled into thinking about social science as an area to be regulated, Gray found himself in an ambiguous position. For lack of expertise, he did not feel able to act as "an inside advocate for social research." Yet he knew more about the social sciences than any of the commissioners, and he did want to advise the commission on the subject.[30] His solution was to try to learn more, but the commission proved reluctant to help. For example, Gray suggested that legal scholar Paul Nejelski ask the commission to support a proposed book on the legal and ethical problems facing social researchers. Nejelski got only a brief note from the commission's staff director, explaining that such a project lay outside the commission's mandate.[31]

Gray kept trying. In the summer of 1976 he wrote to three law professors who had been hired by the commission to report on the legal status of informed consent. He was disappointed that the report seemed to address only medical experimentation, and in five single-spaced pages, he detailed some of the many ways that regulating social science by analogy to medical experimentation might impose the wrong restrictions on investigators. On the one hand, it might be overly protective, depriving citizens of the constitutional right to ask questions. On the other hand, it might overlook threats of invasion of privacy not found in medical research.[32] In response, the lawyers agreed that Gray raised interesting questions. But, they noted, "in our trips to the Commission meetings to present drafts of our papers, *no* Commissioner ever asked us a question about the type of research to which you allude."[33] Gray himself later concluded, "I don't think that we ever got the commission engaged at all in this set of issues. There really wasn't anybody on the commission that cared about them."[34]

Gray had more success with his boss, assistant staff director Barbara Mishkin. Prior to the commission's creation, Mishkin had worked at the National Institutes of Health, where she had helped draft the 1974 human subjects regulations that had been the NIH's unsuccessful attempt to deter Congress from passing the National Research Act. She then helped compile lists of candidates for seats on the commission, before joining its staff herself. But throughout all of this, she focused on biomedical and psychological research until Gray alerted her to the question of the regulation of social science research. As she later explained, "most of what I know about social science research I learned from sitting next to Brad Gray and having him tutor me. We had marvelous discussions about various things, and whether they would or would not fit the definition of research that the commission was struggling to come up with. It was a good education for me."[35]

Despite her eagerness to learn, Mishkin, like the commissioners, seems to have absorbed only half of what Gray hoped to teach. Gray persuaded her that in its potential to invade subjects' privacy, social science bordered on psychology, and therefore might be subject to some of the same restrictions. But he failed to make her see that, as he had put it to the law professors, "the parallels between social research and journalism are, if anything, more striking that those between social and biomedical research," and that social science might be due some of the liberties accorded a free press.[36] Time and again, Gray noted that social researchers often did work indistinguishable from that of journalists. For example, he noted that both sociologists and newspaper reporters had tested housing and lodging discrimination by feigning interest in renting rooms.[37]

The commission never dealt with these challenges. Commissioner Robert Turtle, a lawyer, insisted on "the difference between being a social scientist and an investigative reporter," but could not articulate that difference.[38] Staff director Michael Yesley joked that if social scientists wanted to do research without revealing their agendas, "they should just hire reporters to do it."[39] At best, the commissioners tried to draw a line between privately financed journalism and government-sponsored research. As Ryan put it, "it is one thing when the newspaper sponsors it so that they can write an article. I presume it is not illegal because no one has challenged that. It is another thing when the Federal Government sponsors this kind of research on behalf of the public."[40] But this distinction made little sense, since DHEW had insisted that all university research, regardless of federal sponsorship, needed IRB review. The commission never asked if prior review was appropriate for university newspapers or for reportage by federally sponsored organs such as the Voice of America.

Because of Gray's humble position on the commission staff, he could never force Mishkin or the commissioners to wrestle with the issues he raised. Nor could he persuade them to seek outside expertise. As his fellow staffer, Tom Beauchamp, later recalled, "Brad had a lot of trouble in getting other people to focus on these problems. It wasn't for lack of trying. But he had great difficulty in getting people to focus on anything that was methodological or unique. The commission was looking for general formulae that people could put into practice, particularly in the world of biomedical research."[41]

THE IRB REPORT

The federal law establishing the commission required it to consider "mechanisms for evaluating and monitoring the performance of Institutional Review Boards . . . and appropriate enforcement mechanisms for carrying out their decisions."[42] In fulfilling this mandate, the commission began with a focus on biomedical and behavioral research, and only a stream of complaints by social researchers awoke it to the special problems posed by IRB review of the social sciences. In the end, however, those complaints had little effect.

The problem began with a failure of research. Since 1971, when DHEW insisted that grants and contracts for "an activity involving human subjects" be reviewed by IRBs, universities had expanded their use of such review, including for research not funded by the federal government.[43] But little was known about the effects of such policies, so in June 1975, Gray persuaded the commission to contract for a survey of researchers and IRB members.[44] Conducted between December 1975 and July 1976 by the University of Michigan's Survey Research Center (and approved by that university's IRB), it asked researchers and IRB members to comment on research reviewed between July 1974 and June 1975.[45] It was a massive effort, reaching over 3,900 researchers, subjects, and others at sixty-one institutions.[46] The whole thing cost about $1 million—real money in 1975.[47] The commission received preliminary results in October 1976.

Given the commission's emphasis, the survey predictably focused on biomedical research. The questions about biomedical procedures were fairly detailed. For example, if a researcher reported the "examination of internal structures through natural orifices," he was asked to specify which procedure, and which orifice, he used.[48] The report did cover, in broad terms, behavioral research, but Gray thought that the term meant psychological research—the domain of two of the commissioners.[49] When they wrote the survey questions, Gray and his colleagues were not yet aware that many university IRBs had begun insisting on their jurisdiction over social research. Ironically, it was only after many institutions subjected his survey to IRB review (and some even rejected it) that Gray realized the magnitude of the issue.[50] As a result, the survey team asked almost nothing specific about *social* research outside the controlled settings of laboratory and classroom. The survey itself asked only two questions regarding interviews: whether they were conducted in person or on the telephone, and whether they asked about criminal activity. And it asked only one question about the category of "behavioral observation or experimentation": whether it took place "in a *controlled* setting." The survey team then failed to include even the result of this question in its report to the commission.[51] The questionnaire made no distinction between door-to-door surveys asking political opinions and in-depth oral history interviews, or between watching strangers on the subway and living for years among people of another culture.[52]

The closest the survey came to asking about ethnographic research was a section in which it asked IRB members about six hypothetical cases. In the only nonbiomedical case, "a researcher is doing a paper on the development of institutional policy in which he wishes to interview administrators in the institution." Of the three categories of IRB members surveyed (biomedical members, behavioral members, and other members), majorities of each believed that this hypothetical project should not even be brought before a human subjects review committee.[53] In other words, the only thing the survey learned about IRB review of social research is that most IRB members thought it was inappropriate. But this finding did not make it into published accounts of the survey.

When the survey asked respondents' opinions of IRB review in general—whether it impeded progress, or whether its benefits outweighed the difficulties imposed—it did not differentiate among types of research. Thus a respondent who thought that IRB review was a fine thing for biomedical research but a disaster for sociology had no way of indicating that on the form.[54] This was especially problematic, given the survey's focus on research reviewed between July 1974 and June 1975, a period when most university IRBs likely had yet to intrude on the social sciences. At the University of Colorado, for example, it was not until the spring of 1976 that the IRB forcefully proclaimed its authority over social science research. Therefore, many respondents might have been thinking only of medical research when answering the survey. For similar reasons, the study also failed to reach many social scientists. The survey of roughly eight hundred review board members found only twenty-three sociologists, seven political scientists, four anthropologists, and three economists—not enough to yield representative data.[55]

As a result of this sparse data, table after table in the report presented information about "behavioral research" without distinguishing among such varied approaches as psychological experiments, survey research, and participant observation. The survey also sought information about harms from research, but found little in the nonbiomedical categories. Of all 2,039 projects surveyed, only three reported a breach of confidentiality that had harmed or embarrassed a subject.[56] And of the 729 behavioral projects surveyed, only four reported "harmful effects."[57]

Ignored by the survey, social scientists resorted to writing complaints, which began to arrive at the commission's offices in the spring and summer of 1976. Sociologist Edward Rose, who had kicked up such a fuss at the University of Colorado, complained that his university's Human Research Committee asserted the right to review even research using public information.[58] Some University of South Florida sociologists reported that their IRB had derailed two projects. First, it demanded signed consent forms for what was supposed to be an anonymous mail survey. Second, it requested a list of questions that were to be asked by a researcher planning participant observation, an activity in which questions are not formed in advance.[59] These had so alarmed the department chair that he began to write to other sociology chairs around the country, asking if his was an "isolated instance" or if "the DHEW guidelines on an admittedly appropriate issue [were] being used to control the content of research or limit research methodologies in violation of academic freedom when no real harm to participating subjects can be defined."[60]

Murray Wax, on behalf of the Society for Applied Anthropology, warned that there "was a strong feeling among much of the membership that as some universities have applied and interpreted 'human subjects protection' they are stifling basic and traditional forms of ethnographic and anthropological field researches which had put no one at significant risks, except perhaps the fieldworker himself (or herself)."[61] "Because the process is developmental and organic," he later elaborated, "the project often cannot be initially described, nor can the persons involved offer an abstract consent." He suggested an empirical study of how fieldworkers were faring with their university IRBs—a question Gray's limited IRB survey had failed to ask.[62]

In the fall of 1976, the commission woke up to the problem. At its October meeting, amid a discussion of the IRB survey, staffer Stephen Toulmin noted "a very widespread feeling in the behavioral and social science communities that their problems are not addressed by the present system," while commissioner Robert Cooke conceded that "I received in my institution substantial complaints from the behavioral scientists, from the sociologists particularly, and to some extent psychologists, that much of the HEW guidelines are really applicable to biomedical research and are quite inappropriate for behavioral and sociological research, particularly."[63]

In November 1976, shortly after Congress extended the life of the commission for one year beyond the originally authorized two, a draft report repeated many of the complaints the commission had received from social researchers. It noted that Congress and DHEW had never defined human subjects research and stated that "social scientists inquire about the basis for requiring a signed consent form prior to the conduct of an interview when such a procedures has never been suggested for newspersons whose published work may differ from the social scientist's work only by being quicker, less systematic, less related to theory, and more likely to identify respondents."[64] And, citing Wax, it noted that protocols established in advance and written consent forms made for poor fieldwork.

Similar concerns emerged in a staff paper prepared for the December 1976 commission meeting. The paper, entitled simply "Social Research," outlined forms of social research—observations, questionnaires, and experiments—without trying to gauge the ethical dangers involved or possible solutions to those dangers. On the contrary, it included an excerpt from a paper by two social scientists who argued that while it was possible to "imagine . . . as many ills as possible in order to see what it is, if anything, that needs a cure . . . it is difficult to find much evidence of undeniable outright harm caused to the subjects of social research."[65] The paper helped persuade the commissioners to learn more about social research. At that December meeting, Ryan conceded that "I don't think we can dodge [the issue of social research] . . . and the major question is, in what way is social research different than the biomedical/behavioral kinds of research that we have been dealing with. Can we sort it out?" To answer that question, the commission decided to hold hearings in the spring of 1977, in large part to hear from the critics of IRBs.[66]

The hearings, held in April and May of 1977, attracted dozens of scholars, many of whom arrived with complaints about the review of social science research. The sociologists who testified, as well as a representative of the American Anthropological Association, all complained about the application of medical ethics and methods to nonmedical research.[67] Some critics stressed the rarity of harms from social science research. Paul Kay, the chairman of Berkeley's IRB, complained that "over half the IRB's time is spent on processing the 99% of submissions that pose zero or minimal risk to human beings."[68] Kay also lamented IRBs' zeal for written consent forms, noting that it could actually work against ethical research if research participants felt that signing the form obliged them to keep answering questions.[69] Another witness argued that "voluntary participation of the subjects can be assured by several methods, not just rigid adherence to a written consent form."[70] Sociologist John Clausen noted that some adults who would be happy to talk were nevertheless suspicious of forms. He pled that "competent adults be treated as competent, be allowed to decide whether they want to participate in an interview without necessarily having to sign a form."[71]

The critics also argued that it was foolish to apply rules for experiments—which by their nature begin with fixed questions and protocols—to the more fluid work of social research. One witness, herself a chemist, warned that "the extent to which the research can be designed and explained in advance is different" in social science, and that it involved "continually evolving research designs."[72] Another critic added that "the field work and participant observation methodologies of social science and the curriculum development work of educational research often have gradually evolving research designs rather than formal experimental designs used in bio-medical and psychological research."[73] Hans Mauksch, the executive officer of the American Sociological Association, argued that an IRB's main tool—prior review of research protocols—was all but useless in protecting participants in social research. "The very nature of some qualitative research precludes the specific documentation of every question that will be asked," he explained, "where the research requires learning from the subject which questions need to be asked and what methodology should be followed and adopted." When sociologists did engage in ethically risky research, such as interviewing dying persons, "the real issue of the protection of the human subject has to be placed on the ethics, the competence and the sensitivity of those professionals who are conducting the research."[74]

The one form of social research that can most easily be vetted in advance—survey research—faced its own problems. As Ronald Heilmann explained, the DHEW guidelines required a subject to be given six specific elements of information, such as "a description of any attendant discomforts and risk" and "a disclosure of any appropriate alternative procedures." Important to medical experimentation, these requirements were irrelevant to much survey research. Adding them to a survey researcher's script would only bore or scare away potential respondents, damaging the sample.[75]

In both correspondence and at the hearings, the critics proposed alternatives to IRB review. Wax conceded that there might be a role for review—not before a project began, but as it evolved and was completed. He also pointed out that anyone who felt hurt by a publication could sue for invasion of privacy or libel.[76] Others mooted the idea of a review of graduate student research within academic departments, where, presumably, faculty would be more familiar with the methods of social scientists than would boards drawn from other disciplines.[77] Similarly, one sociologist suggested that only IRBs that included experienced ethnographers were qualified to judge proposals for field work.[78] Herbert Phillips of the University of California, Berkeley described the plan proposed at that institution to allow researchers to decide if subjects were at risk and, if they were not, to file an affidavit with their department rather than submit the project to IRB review. Since DHEW had killed that plan, Phillips pleaded with the commission to overrule the department.[79] (A 1978 paper written for the commission found widespread support for the affidavit system among experts on social science.)[80]

The commissioners and their consultants heard the complaints—to a degree. Brady described the events at the University of Colorado as "IRB atrocities" and noted "a lot of testimony from social scientists saying that the procedure is causing one sort or another problem in the review of social research." Turtle fretted that IRBs were always taking the safe route of being restrictive. And Levine, who two years earlier had suggested that social science was not fundamentally different from biomedical research, now noted "that a lot of [researchers'] problems had to do with applying the model either of biomedical research, or the modes of thinking that are derived from the 'hard' sciences, from biology and chemistry and things like that, to the social sciences."[81]

Following the 1977 hearings, the staff prepared a new draft of what would become the IRB report. This draft noted that "the most frequent general criticism of the current activities of IRBs is that they operate under regulations developed for one type of research (biomedical) that are disfunctional [*sic*] when applied to some other types of research (e.g., educational research, epidemiological studies of medical records, survey research, observational research, ethnography and sociolinguistics, and evaluation research carried out in connection with social service programs)." And it presented ways of keeping some oversight of less-risky research while avoiding full IRB review: the Berkeley system of affidavits, departmental committees, and preliminary review by IRB subcommittees or staff.[82] Along with the report, the staff submitted an eight-page list of "Issues Relating to the Performance of Institutional Review Boards," asking, among other things, whether IRB review was appropriate for linguistics, folklore, ethnomusicology, oral history, surveys, or studies of publicly available information.[83]

Most significantly, the report raised the question of how to define a human subject—a definition missing from both the 1974 regulations and the National Research Act. It offered three suggestions for defining human subjects in such a way that the definition would include all the activities investigated by the Senate in 1973 while excluding the social sciences, which had not been the Senate's concern. The first option would be a specific list of exempt activities. The staff report contained eleven suggested categories, including several that would exempt social research. One suggestion would be to exclude "linguistics, folklore, ethnomusicology, oral history, and ethnography," as well as "studies of publicly available information." Perhaps even broader was a suggestion to exclude the "wide range of research, training, instructional, and demonstration activities which occur on a daily basis within a university." The second option would be to define a human subject as " 'a person who is under the immediate control of and whose behavior is being modified by the commands of a research investigator,' thereby excluding observational research, studies of records, and most survey research." And the third option would be to define risk "(a) realistically (so that, e.g., an IRB would not judge the act of asking a potential subject to participate in a survey to constitute a risk) and (b) so as not to include risks to social categories of which subjects are members."[84]

The commission, however, proved unable to agree on a definition of human subjects research. At the July 1977 meeting, commissioner Albert Jonsen suggested defining a subject as "one who falls within the power of an investigator such as that person's life, health, well-being, reputation, dot, dot, dot, whatever other qualities we consider important can be adversely affected and who cannot, for some reason, defend themselves [*sic*]."[85] This was not terribly far from the definition offered by the OPRR's Chalkley in the Georgia Medicaid case, in which he had testified that a human subject was someone "who is deliberately and personally imposed upon."[86] But Lebacqz was troubled about "the part about the people not being able to defend themselves." And Mishkin and Levine worried that it would be too hard to determine how much power the researcher has; they felt that such a determination would be best left to an IRB. Levine suggested that it would be better to have a definition that was too expansive, since each IRB would "take very little interest in" some projects that came under its jurisdiction.[87] The conversation moved on without a commission decision on a definition of human subjects research.

This left the decision—one of the most important made by the commission—back in the hands of the staff. Led by Mishkin, the staff produced an extremely expansive definition: "A human subject of research is a person about whom an investigator conducting research obtains (1) data through intervention or interaction with the person, or (2) private information. 'Research' is a formal investigation designed to develop or contribute to generalizable knowledge. 'Interaction' refers to communication or interpersonal contact between investigator and subject."[88] This definition abandoned both the DHEW regulations' focus on "subjects at risk" and the congressional concern with "biomedical and behavioral research," the words embedded in the commission's title. Instead, it could include all surveys, all fieldwork, even all conversation.[89]

The staff members who crafted this definition did not intend to regulate some of the activities that IRBs would later monitor. Asked later if she had intended to regulate journalism, Mishkin replied:

> The commission, probably not. Brad, probably not. I, probably not. It wasn't brought up often. It was brought up from time to time in discussions, and usually in a context in which the example was clearly not within the definition of biomedical and behavioral research that we were working with. It was not something that was an organized activity designed to contribute to generalizable knowledge, by which they had in mind generalizable knowledge in a scientific field. Which is different from journalists, obviously, who want to contribute to the knowledge of the general public. I do not think that is what the commission meant when they said contribute to generalizable knowledge.

But the definition itself offered no such guidance; every newspaper article, and every call to a government office, would conceivably qualify.

Other staffers and consultants understood the problem. At the next commission meeting, Levine warned that "in order to deal with the definition as drafted here, it becomes necessary then to write an awful lot of exceptions and 'but we really don't mean this, and we really do mean that.' "[90] Gray later explained that he was uncomfortable with the broadness of the definition but saw no alternative. "If we could have come up with an argument that we believed, that said, there is a principled basis for excluding social science from this whole framework that is designed to protect the rights and welfare of human research subjects, I would guess that most of us would have grabbed it," he recalled. "I would have . . . I felt like we were stuck in including social research within the boundaries of research involving human subjects."[91] But the commissioners failed even to grasp the dilemma, and at their August meeting they adopted the definition with no discussion of its effect on the social sciences.[92]

Then, in September, the staff presented a draft list of recommendations, most of which would remain in the final report. It suggested that every federal agency, and every institution that received federal funds, should establish IRBs, which, in turn, would have the power to review all research involving human subjects, whether federally funded or not. IRBs would be empowered to modify, disapprove, monitor, and suspend research, and to report noncompliant investigators to "institutional authorities."[93] Noncompliant institutions could be deemed ineligible for all federal funding.

Amid these draconian measures, Gray secured three key concessions for social scientists. First, as Gray had argued for some time, the report stated that "the

IRB should . . . not consider as risks the possible consequences of application of the knowledge gained in the research (e.g., the possible effects of the research on public policy.)"[94] In theory, at least, this would allow social scientists to criticize organizations, or ethnic groups, so long as they did not harm individuals. Thus it resolved the issue debated at Berkeley in 1972 and 1973 in favor of researchers. Second, the report did not require the signed consent forms typical of medical research to be applied in all cases. Instead, it allowed IRBs to waive the general requirement for written consent either when a signed consent form might place the subject at risk or when obtaining it "would impose burdens upon the research or the investigator that are not justified by the increment of protection that written consent might provide," such as in "a telephone survey or an ethnographic study."[95]

Finally, the draft recommended new "expedited review procedures, short of regular IRB review, for carefully defined categories of research that present no more than minimal risk." These categories would require approval from the "responsible federal agency or department," though the draft failed to explain which agency would be responsible for research not funded by any federal agency. The draft then listed categories that "might be suitable for expedited review." Most were procedures—analysis of hair and nail clippings, small blood samples—suggested for exemption by the Clinical Center of the National Institutes of Health the previous year.[96] But the draft also placed on the list the use of "survey research instruments (interviews or questionnaires) . . . provided that the data will be either gathered anonymously or will be protected by confidentiality procedures appropriate to the sensitivity of the data." Such research would be held to the same ethical standards as other research, but it could be approved by an IRB chairperson or a designated IRB member or staff member.[97]

The recommendation for expedited review was a leap into the unknown. Approval (or rejection) of a project by anything less than a full IRB was forbidden by the regulations, and the 1976 IRB survey conducted for the commission noted that in all the institutions it studied, "individual reviewers were never reported to make decisions for the committee regarding a proposal's acceptability."[98] (Informally, IRB chairs did approve projects at some universities.)[99] All that Gray could say in favor was that boards that had a special review of some projects by individuals or subcommittees in advance of full board review performed better on some measurements, including investigator satisfaction.[100] But compared to the alternatives considered—the filing of affidavits or a review by departmental committees—review by one IRB member was the least respectful of disciplinary differences. A sociologist filing an affidavit could be expected to know her own research ethics, and an anthropologist would share an ethical and methodological framework with members of his own department. But a researcher applying for expedited review had no guarantee that the designated IRB member or staffer would have any acquaintance with the type of research proposed.

Yet even at this late stage—in the fall of 1977—the commissioners seemed barely aware of the impact their recommendations could have on the social sciences, despite the efforts of staff and consultants to interest them in that problem. Joel Mangel, a DHEW lawyer, begged for clarification of whether the recommendations were designed to cover social science and policy experiments as well as biomedical and behavioral research. "If it is intended to cover that kind of an activity, I think it would be helpful to say so, and if not, also to say so." Gray told the commissioners—as if for the first time—that many social scientists considered the design of the IRB process to have been "developed on the medical model" and some of the requirements to be "silly."[101] And consultant John Robertson brought to the commission's attention a letter from the University of Illinois—which had a well-established system of departmental review for unfunded, low-risk projects—complaining that abolishing that system "would require very substantial increases in time, effort, and funds with *no gain* in protection of subjects."[102] None of these pleas provoked a response. The commissioners did not even bother to write comments on the staff recommendations.[103]

Only in December 1977—seven months after the hearings—did the commissioners acknowledge the problem. Donald Seldin suggested that "a different organizational structure, different expertise, and certain different local guidelines be applied where social science research is the subject matter of the IRB."[104] Albert Jonsen opined that social research presented "a very different kind of risk, a very different perception of a problem than the one we usually think of in the biomedical world." Chairman Ryan acknowledged that sprinkling the recommendations with the qualifier "where appropriate" would not be enough, and that the report needed language to "indicate our awareness that some forms of behavioral and social science research and so on are going to have to be looked at with respect to their needs rather than strictly on a biomedical model." But they offered no specific provisions for ensuring that social research would be reviewed by boards that understood that research. Nor, in their debates over various recommendations, did they discuss how those recommendations might affect social science. For example, when the commissioners decided to reject the requirement in the 1974 regulations that benefits outweigh risks in all cases, they did so after discussing a variety of medical cases—real and hypothetical—with no one mentioning the social scientists' frustrations with the risk-benefit framework as a whole.[105]

The commissioners also declined to offer scholars much protection against capricious IRBs. Gray warned the commissioners that researchers had complained of boards "constituted to evaluate one kind of research" making uninformed judgments about the risks of another kind of study. He cited a paper arguing that people not familiar with deceptive research in social psychology "frequently vastly over-estimate the extent to which it is traumatic to subjects."[106] But the other staffers and commissioners decided that subjecting IRB decisions to appeal would undermine IRB authority. Mishkin was sure that a proper IRB would include members familiar with every kind of research it reviewed.[107] Consultant Stephen Toulmin was particularly dismissive of concerns about the rights of researchers, noting "the question of whether we are going to regard this thing as an abuse of academic freedom is a question which is clearly going to have to be argued out between the [American Association of University Professors] and the university administration. I mean, it is not something, it seems to me, which falls within the Commission's terms of reference here." [108] Jonsen later boasted that, compared to the existing 1974 regulations, the commission's regulations offered "a great and broad flexibility" for IRBs to review "any sort of research that came before them."[109] But the commission offered no guarantee that the boards would use that flexibility wisely.

By February 1978, with the IRB report almost finalized, it was left to Gray to warn that some agencies, such as the Office of Education, legitimately resisted the DHEW regulations

> because they believe that the regulations are not appropriate to the research that they conduct. The Commission has not looked at that research. We are saying that these regulations should apply, but we have not looked in any detail at educational research. There are a number of things that agencies do that could easily be construed to be research involving human subjects that we have not even thought about.

His superior, Barbara Mishkin, disagreed. "I think we have built in enough flexibility in these IRB recommendations to accommodate any social research, any social science research, any research in education, and so forth," she opined, and the commissioners moved on.[110]

As the commission completed its work in the spring of 1978, Gray tried again. He asked commissioner Turtle why a sociologist should have less leeway than an investigative reporter, but got no response. He argued that sociologists should have the right, "as anybody else has the right," to film public behavior, such as demonstrations. Yesley, the commission staff director, replied that researchers in institutions submitted themselves to IRB jurisdiction when they signed their contracts. Gray retorted, "That wasn't in my contract when I taught."[111] Chairman Kenneth Ryan, perhaps distressed by the disagreement, noted, "We spent a whole morning talking about [observational research], and obviously there has been a lot of discussion, so that there is a difference of opinion and confusion about the subject. The commission, over its long time, has not really had an opportunity to spend adequate time on the social science research problem."[112]

The commission never found that opportunity, so it contented itself with making small adjustments to its recommendations. Gray gained the concession that "informed consent is unnecessary . . . in studies of public behavior where the research presents no more than minimal risk, is unlikely to cause embarrassment, and has scientific merit."[113] But left unresolved was the broader question of why university researchers should require IRB permission to do things that journalists and private citizens did all the time. All the social scientists got in this regard was an admission in the final report that the commission's expansive definition departed from tradition. The report noted that when it came to research by federal agencies themselves, "questionnaires and surveys" were "activities about which there is presently no uniform understanding with respect to the nature and extent of protective mechanisms that should be applied." It went on to note that although

> survey research entailing no intervention in the lives or activities of the subjects [falls] within the Commission's definition of research with human subjects, it should be noted that data gathering, in and of itself, has not universally been considered "research with human subjects."[114]

The commission admitted a lack of consensus about whether surveys even fell under the rubric of human subjects research and acknowledged the disagreement over whether an IRB was the right mechanism to oversee them.

Nonetheless, the commission's doubts did not deter it from imposing IRB review on a broad range of research. In September 1978 it released its final IRB report, descriptively if not elegantly titled "Institutional Review Boards: Report and Recommendations of the National Commission for the Protection of Human Subjects of Biomedical and Behavioral Research," and published it in the *Federal Register* in November of that year. The report formally recommended the staff definition of a human subject: "a person about whom an investigator (professional or student) conducting scientific research obtains (1) data through intervention or interaction with the person, or (2) identifiable private information." It then recommended that "each institution which sponsors or conducts research involving human subjects that is supported by any federal department or agency" be required to assure that "all research involving human subjects sponsored or conducted by such institution . . . will be reviewed by and conducted in accordance with the determinations of a review board."[115] In short, the commission recommended that the federal government require IRB approval of all interview, survey, and observational research at every university that took a single dime from the federal government.

To justify this recommendation, the final report papered over the complaints of social scientists that had poured in throughout the commission's existence. The report claimed that

> IRB members and investigators were virtually unanimous in agreeing that the IRBs at their institutions help to protect the rights and welfare of human subjects, and most agreed that the procedures are reasonably efficient and even that they have had the effect of improving the scientific quality of research. There are some serious criticisms of IRBs as well, particularly from among social and behavioral researchers. Nonetheless, researchers as well as IRB members seem to recognize the need for the review of research, to accept the legitimacy of IRBs, and to be prepared to play a role in supporting the work of IRBs.[116]

These were doubtful claims. The same 1975–1976 survey that underlay them also showed that nearly half of the responding behavioral and social researchers felt that their institutional IRB "gets into areas which are not appropriate to its function" and "makes judgments that it is not qualified to make," and that a majority of those researchers felt that IRBs had "impeded the progress of research" at their institutions.

Worse yet, the conclusion privileged that survey—which had asked no questions specific to social science research—while ignoring the testimony of social scientists at the 1977 hearings, the only chance the commission had given sociologists and anthropologists to make their case. (Just this sort of privileging of quantitative over qualitative evidence would later become the source of many qualitative researchers' complaints about IRBs composed of quantitative scientists.) In their testimony at the hearings—as well as the materials they sent in over the transom—social scientists had made it quite clear that they saw no need for a review of their research, that they rejected IRBs as illegitimate, and that they were doing all they could to eliminate IRB jurisdiction over their work. But in its IRB report, as in so much of the work that led up to it, the commission simply ignored experts in the social sciences.

Over the course of four years, the National Commission's members, staffers, and consultants did an astounding amount of work, gathering data and offering recommendations on some of the most difficult ethical questions then facing biomedical and psychological researchers. But in their work on IRBs, they failed in their responsibilities to social science and to social scientists in two ways. First, they failed to define the term "behavioral research" that comprised part of the commission's own title. The commissioners and staff rightly recognized the importance of definitions, especially when Congress had neglected to provide them. As Levine put it in April 1975, "we can use words all over the place and unless they have a meaning that is shared by all of us all the sentence structure around the words has no meaning understood by any of us."[117] In the course of its work, the commission defined everything from "research" to "human subject" to "justice" to "biomedical." Yet it ignored explicit pleas for a definition of "behavioral research" and for clear boundaries among behavioral research and social science, the humanities, and journalism. Lacking a definition, one commission consultant, Don Gallant, believed that behavioral research meant the quest for therapy or the etiology of a behavioral disorder; consultant Albert Reiss thought the most typical form of behavioral research was a sample survey; while Murray Wax kept asking about fieldwork.[118] It was as if a national commission on the lime industry had completed its work without deciding whether it was regulating citrus fruit or calcium oxide.

Second, the commission failed to investigate what harms and wrongs, if any, social scientists were committing, how IRBs already handled various types of social science research, and what alternative remedies existed. The lack of a definition of behavioral research, and the dearth of early attention to social science, led to the inadequate sampling in the Michigan survey, leaving both IRB advocates and critics to rely on anecdotal argument and speculation about the effects of IRBs on surveys, observations, and interviews. Nor did the commission explore alternatives to IRBs, as social scientists begged it to do.

The commissioners were not wholly deaf to these concerns. In commission deliberations, the commission members and staff spoke openly about their failure to investigate the effect of IRBs on the social sciences and their misgivings about the applicability of some of their recommendations. Yet rather than acknowledge those uncertainties and call for more study, the commission's official reports asserted that social science required oversight and that IRBs were the best means to assure such oversight. In the end, the commission published conclusions about IRBs that were unsupported, or even contradicted, by the evidence the commission had gathered. Similar fuzziness about the social sciences left flaws in the National Commission's most famous product, the *Belmont Report.*

CHAPTER FOUR

THE *BELMONT REPORT*

The National Commission's IRB report remains the basis for much policy on human subjects research, but three decades after it completed its work, the commission was best remembered for another report, entitled *The Belmont Report: Ethical Principles and Guidelines for the Protection of Human Subjects of Research.* The report now has quasi-legal force. Nearly every university IRB in the United States submits a Federalwide Assurance that pledges that "all of the Institution's human subjects research activities, regardless of whether the research is subject to federal regulations" will be guided either by the *Belmont Report* or by "other appropriate ethical standards" certified by the federal government.[1] In practice, almost all American universities agree to follow the *Belmont Report.*[2]

Having so pledged, IRBs treat the report as the central source of ethical judgment. In many cases, they require every researcher to read the report and accompanying commentary. And when IRBs deliberate, they typically measure a proposal against the report's standards and guidelines.[3] As one commentator wrote in 2006, the report "continues to wield totemic influence over the practice of research ethics . . . It runs about the same length as the U.S. Constitution; its students frequently cite 'original-intent' when interpreting the report's ambiguous passages; and . . . its pronouncements command a similar respect."[4] In fact, the *Belmont Report* is even more totemic than the Constitution in that it cannot be amended—it will always remain what it was in 1978.[5] In that sense, at least, the *Belmont Report* is more like the Ten Commandments: fixed in stone.

But the report is the product of fallible mortals. At the time of its twenty-fifth anniversary, bioethicists debated the *Belmont Report*'s utility, and several of them (including some of the authors of the original report) found significant flaws in it.[6] But whatever its merits or demerits as a guide for medical experimentation, the *Belmont Report* is a poor guide for the social sciences and humanities. As Mary

Simmerling, Brian Schwegler, Joan E. Sieber, and James Lindgren noted in 2007, the report "is based on the medical model of decision-making" and "some of the current confusion in the regulation of research ethics, reflected prominently in *Belmont*, stems not only from the field's medical origins, but also from the confusion of research participants with patients."[7] That confusion was the product of a process every bit as narrow-minded as the one that produced the National Commission's IRB report. Like the authors of that report, the commissioners and staffers who wrote the *Belmont Report* were willing neither to listen to social scientists nor to leave them alone.

PRINCIPLES

In the National Research Act, Congress had asked the commission to "identify the basic ethical principles which should underlie the conduct of biomedical and behavioral research with human subjects."[8] But the commissioners began this task by addressing purely medical contexts. They first debated the principles in early 1975, during work on research on the fetus—a medical debate. Issued in July 1975, that report stated that "freedom of inquiry and the social benefits derived therefrom, as well as protection of the individual, are valued highly and are to be encouraged. For the most part, they are compatible pursuits. When occasionally they appear to be in conflict, efforts must be made through public deliberation to effect a resolution." It then presented three principles that would guide such a resolution: "(1) to avoid harm whenever possible, or at least to minimize harm; (2) to provide for fair treatment by avoiding discrimination between classes or among members of the same class; and (3) to respect the integrity of human subjects by requiring informed consent." Yet the report labeled these as "interim" principles, as the commission had not yet begun the core task of identifying principles.[9]

That work began in late 1975, as the commission received background papers by consultants. Besides some rather abstract papers on the nature of ethical principles, the commission received papers specific to human experimentation from Tristam Engelhardt and LeRoy Walters. Both were affiliated with biomedical institutions—Engelhardt with the University of Texas Medical Branch and Walters with Georgetown University's Center for Bioethics. Both produced studies that stayed close to the congressional language of "biomedical and behavioral research." And both relied heavily on three documents derived from biomedical debates: the Nuremberg Code of 1947, the Declaration of Helsinki of 1964, and the existing regulations of the Department of Health, Education, and Welfare, codified in 1974. Walters acknowledged that these were biomedical sources, but he hoped that "many if not all of the ethical principles developed in the essay are also applicable to behavioral-research activities," which he did not define.[10] Engelhardt, in contrast, addressed the behavioral sphere by consulting the American Psychological Association's *Ethical Principles in the Conduct of Research with Human Participants*. Neither scholar mentioned ethics developed in anthropology, sociology, or any other social science.

From their biomedical and behavioral sources, the two scholars abstracted moral principles. Engelhardt emphasized three: "respect for persons as free moral agents, concern to support the best interests of human subjects in research, and interest in assuring that the use of human subjects in experimentation will on the sum redound to the benefit of society."[11] Walters, for his part, offered "four general requirements for ethically acceptable non-therapeutic research: adequate research design; a favorable risk-benefit ratio; equitable selection of subjects; and a reasonably free and adequately informed consent by the subjects."

An anonymous staffer soon realized the difficulty with relying just on the medical codes and offered the following warning:

> A general exhortation merely to 'minimize the risk/benefit ratio' smacks too much of apple pie: there are qualitative differences between the particular kinds of 'harm' and 'benefit' at issue in, say, experimental psychosurgery, first-stage drug testing, psychological experimentation, and social survey research, and their precise ethical significance has not yet been sufficiently analysed.[12]

Similarly, Richard Tropp, who had served in the Office of the Secretary at DHEW, warned that the DHEW guidelines—treated by both Engelhardt and Walters as major sources for ethical wisdom—had been drafted hastily by the National Institutes of Health without consulting the nonbiomedical agencies within the department. "Based on the conceptual framework of a biomedical research model," he wrote, "the current regulation on protection of human subjects is inappropriate, in a number of major respects, to effective regulation of social science research."[13]

The two social scientists asked to prepare background papers split over the issue of applying medical ethics, in part based on their own experiences as researchers. Sociologist Bernard Barber of Barnard College was an expert in the sociology of medical research, having recently coauthored a book about medical experimentation involving humans. As part of that work, Barber had surveyed IRBs but deliberately asked only about biomedical research, thus depriving himself of any insight about the problems faced by nonmedical researchers.[14] Yet

Barber saw no problem in applying medical ethics to nonmedical studies. At Columbia University, he chaired the Human Subjects Review Committee, an IRB devoted to nonbiomedical research. For his paper, he examined three years' worth of his board's files and found that nearly a third of the reviews had flagged the potential for some kind of social injury, such as "embarrassment, loss of privacy, disclosure of confidential information, danger of arrest, adverse effects on family or larger social network relationships, anxiety, fear, self-incrimination, and harmful new self-awareness."[15] Barber did not explain what kinds of research came under his committee's jurisdiction, and instead presented all "behavioral research" as a single category. Nor did he explain how the identification of potential harms to subjects affected the research under review or report the contempt many Columbia faculty held for his IRB.[16]

Sociologist Albert Reiss of Yale came from a more critical tradition. In the late 1960s, for example, he had observed police officers after telling them that his study "concerned *only*... citizen behavior toward the police and the kinds of problems citizens make for the police." In fact, Reiss and his coauthor were equally interested in police behavior. "In this sense," they conceded, "the study involved systematic deception."[17] Unsurprisingly, Reiss was more skeptical than Barber of the existing DHEW regulations. Though he did not explicitly condemn IRB jurisdiction over social science—he seems have taken that as a fait accompli—he did suggest that the biomedical model encoded in the regulations was inappropriate for much social research. For one thing, it assumed that investigators were in full control, something much more true in a medical clinic than in (say) a telephone survey, where the subject could always hang up. Second, it assumed that the investigator began with a fixed list of procedures, hardly the case in much of social science. Third, it assumed that all harms were to be minimized, not the case in "muckraking sociology or social criticism." Overall, he found, "behavioral science inquiry is generally low risk inquiry so that for much of it a requirement of informed consent seems unnecessary and burdensome." If governments really wanted to protect participants in such research, they should offer "a legal privilege against compelled disclosure and . . . legal penalties for unauthorized disclosure, misuse, or illegal use."[18] But Reiss buried these points in a rambling, 164-page essay that lacked the passionate clarity of earlier sociologists' critiques of IRBs, such as Edward Shils's six-page attack on the Berkeley policy of 1972.[19] The case against universally applying medical ethics lacked an eloquent spokesman.

The commissioners began to debate such issues when they met in February 1976 at Belmont House, a conference center in Elkridge, Maryland—a meeting that was to give the *Belmont Report* its name. Some favored IRB review, and the potential for restrictions, for research that consisted only of asking questions. Karen Lebacqz noted her concern for "native American peoples, who are studied by anthropologists continually," and that "there has been a certain amount of public consternation about behavioral research, possible abuses of persons or groups of persons arising from behavioral research," though she did not give examples of the latter.[20] Robert Cooke, calling for IRB review even of observational studies, argued that "you just do not turn people loose that might use information against individuals and so forth."[21] And Turtle noted that he was "troubled by the acceptance of the fact that anybody can go out and ask anybody else questions, at least within the context of a research situation, that somehow there is no harm or risk, and that we ought not to be terribly concerned about it, and that it is somehow different than other types of research which I think we look at in the biomedical sense, and say that they create risk."[22]

In contrast, commission staff and consultants called for a distinction between biomedical ethics and ethics for social and behavioral sciences. Staffer Stephen Toulmin opined:

> One of the questions that is obviously going to have to be looked at very carefully when we move on to the IRB examination phase is simply the question of whether this same set of procedures can meaningfully and effectively be employed for the appraisal of biomedical research involving human subjects and these other kinds of research, especially in the field of social science.[23]

Two consultants shared his skepticism, offering cases where substantial harm to subjects might well be ethical. Reiss noted that Abraham Flexner's 1910 research on medical education, considered a milestone in the professionalization of medicine, could well have been considered harmful research because it resulted in the closing of dozens of shaky medical schools.[24] And law professor Charles Fried insisted that "freedom of inquiry" should be one of the core ethical principles of research. He proposed a hypothetical case in which a researcher would be restricted from analyzing the thoughts of federal judges as revealed in their published opinions. "It may be that those judges are under extreme risk that what might happen to them would be terrible," he explained, but "I take it that they have no right to object, that nobody has a right to tell you you cannot do that, that that is prior restraint; it is unconstitutional; it is censorship; it is horrible."[25]

Yet this debate did little to shape the report on ethical principles. A discussion outline used at the Belmont meeting, for example, introduced its list of ethical principles by referring to "the accepted codes on human experimentation," which presumably meant the Nuremberg, Helsinki, and DHEW guidelines. Following Walters's precedent, it suggested that "relevant sections of the codes will be quoted" for each principle. Indeed, Toulmin's first draft of the ethical principles report, dated 1 March 1976, built on the work of Engelhardt and Walters by boiling down the Nuremberg, Helsinki, and DHEW codes into "fundamental ethical requirements of three kinds: viz., respect for persons, justice, and beneficence." Respect for persons included concerns about the freedom to choose whether to participate in research and provisions for informed consent. Justice, in this case, corresponded closely to Walters's concern for the "equitable selection of subjects." And beneficence combined Engelhardt's "interest in assuring that the use of human subjects in experimentation will on the sum redound to the benefit of society" with Walters's wish for "a favorable risk-benefit ratio."[26]

Lebacqz acknowledged that the February outline "might have been a little heavily weighted towards the biomedical side," so she attempted to include "concerns for behavioral research" as well in a March draft.[27] But the results of this effort were two provisions, included in a new section on "Norms for the Conduct of Research," that suggested heavy IRB intrusion into social research. The draft argued that applying the principle of beneficence meant that "it is necessary to take into account, not only the effects of the research on the individual research subjects directly involved, but also on any others who may be involved indirectly (e.g., in social research that investigates the institutions in which they work) and also any broader classes of persons whose interests may be adversely affected by the research."[28] In other words, the IRBs would have to determine, in advance, the effect of a sociological or anthropological study on people who did not take part in the study, a determination that, as Rciss had warned, was impossible, and one that had pushed the Berkeley faculty into rebellion.[29]

Another part of the "Norms" section warned that "the selection of subjects for social science projects, particularly, the use of the so-called 'snowball' technique calls for special ethical care. If an interview with one research subject is to be used as a means of recruiting further research subjects, there is a significant risk of failing in respect for the integrity of the original subject, and using him 'not as an end in himself, but as a means only.' " The snowball sample, an established technique of social research, simply meant asking participants whom else the researchers should talk to. It had been presented, with no hint of censure, in a widely used textbook by the director of the same University of Michigan institute that had conducted the IRB survey for the commission.[30] The draft thus advocated an application of Kantian principles so strict that it forbade the everyday tools of social science.

The March 1976 draft languished for the rest of the year, with none of the commissioners nor their staff confident about revising it. Consultant Stephen Toulmin, who had originally been in charge of drafting the report, was pulled away on other duties.[31] So in December, staff director Michael Yesley hired philosopher Tom Beauchamp to revise the report.

APPLICATIONS

From his arrival in December 1976 through the remainder of the commission's existence, Tom Beauchamp was the only staff member for whom the Belmont paper was the top priority. At the time of his hiring, Beauchamp was a rising authority in the field of biomedical ethics and was at work coauthoring a book later published as *Principles of Biomedical Ethics*.[32] But like the commissioners themselves and the majority of the staff, Beauchamp knew relatively little about the ethics of social science research. As a result, his work contributed to the report's evolution into a manifesto of medical ethics.

Beauchamp arrived at the commission believing that its task was to right the wrongs of medical research. As he later recalled, in

> the 1966, 1967, 1968 period, when all of this gets underway about IRBs, and worries about ethics, and so on, it's entirely NIH oriented, at its origins. Entirely. To suppose that by the time the National Commission came along and the world "behavioral" got stuck along[side] "biomedical," that that history would be gone, would be foolish.

Medical research was clearly having ethical problems; Beauchamp saw his job as finding ways to restrain doctors. As he put it, "we're coming from a situation in which the judgment had clearly been made that there was too much freedom on the part of the scientific investigator. Too much freedom not to attend to moral matters. That was the problem." For this reason, he scorned Fried's idea that "freedom of inquiry" might be a fundamental principle.[33]

Beauchamp was aware that the word "behavioral" was in the commission's title, and he made some effort to study the problems of nonmedical research. But in doing so, he concentrated on the ethics of psychologists, for two reasons. First, as Beauchamp himself pointed out, "two commissioners were psychologists, so we wanted to be sure to have those bases covered." Had the commission included members of all the scholarly disciplines it sought to regulate, those members could have insisted that the *Belmont Report* reflect their ethics. But, recalled Beauchamp, when it came to the social sciences, "there was virtually zero expertise and/or interest among the commissioners in this issue."[34]

The second reason that Beauchamp focused on psychology was that he believed it had better-developed ethical guidelines than did other behavioral and social sciences. In particular, the American Psychological Association (APA) had recently published its 104-page book, *Ethical Principles in the Conduct of Research with Human Participants*, offering both general principles and commentaries on specific cases.[35] In contrast, Beauchamp later disparaged the ethical codes put forward by sociological and anthropological organizations as "extremely thin, like it could be condensed to a single sheet of ten principles, or something like that. Pretty uninteresting."

Was this comparison fair? Those disciplines had indeed tried to reduce guidelines to short, memorable lists, but so had the American Psychological Association, whose book began with a list of ten ethical principles of just the sort that Beauchamp disdained.[36] And if no other discipline had published an official, book-length examination of research ethics, anthropologists and sociologists, in particular, had written extensively about the ethical challenges of their work. For example, sociologist Myron Glazer's 1972 book, *The Research Adventure*, can be read as sociology's unofficial equivalent of the APA publication.

Most likely Beauchamp read none of this literature and learned little about social science ethics from any other source. Interviewed in 2007, he recalled no significant engagement with the papers written in preparation for the Belmont meeting, such as Reiss's rambling explanation of sociological methods and ethics. He did not recall that he (or anyone else except for Bradford Gray) was influenced by the critiques social scientists, such as Murray Wax, sent to the commission. Nor was he affected much by the IRB hearings of 1977; he had not attended them. However, *Tearoom Trade* had made enough of an impression on Beauchamp that he brought it up in a commission meeting; he took it as evidence that social scientists had too much freedom.[37]

Nor was Beauchamp striving to encode the standards already held by investigators. As he later explained:

> A principle means, not what is believed by some group of people, or believed universally, but what is normatively correct. It is a normative principle, what is in the federal law is called the principles that should govern research. *Should* govern. Not have governed, have been respected in the past.[38]

Whatever social scientists—or any other group of researchers—had done, and whatever they had said and written about what they had done, were irrelevant. What mattered to Beauchamp was what they *ought* to have said and done. As he noted in December 1977, "I think this is a paternalistic commission. It out and out is a paternalistic commission, at least in the way I understand the issue of paternalism."[39]

On the other hand, as he conceded in 2007, the commission's deep lack of expertise was a problem. "You cannot do good work in [professional ethics] unless you have a pretty good understanding of the area that you're concerned with . . . For example, if you're into the ethics of genetics, if you don't understand genetics, you can't do it. And so on. Did we do that [good work] on the commission when it came to the social sciences? Absolutely not."[40]

Beauchamp's lack of expertise in the social sciences was significant, because when he began work the ethics paper was still fluid, as noted by a February 1977 staff paper entitled "Some Issues Not Fully Resolved at Belmont." The paper explained that the principles of respect for persons and beneficence came into conflict when one person judged that another person should not be allowed to take part in risky research. Autonomy should trump beneficence, it argued, so "IRBs should not decide against a research project on grounds that (competent, consenting) persons would place themselves at too much risk." It also rejected the March 1976 proposal that beneficence had to include consideration of the ramifications of research findings. The February 1977 paper responded that "it has never been part of the ethos of research, scientific or otherwise, that research should be done only if there is a reasonable assurance that the findings will not cast any segment of society in an unfavorable light."[41]

In April 1977, Beauchamp presented the commission with a revised Belmont Paper, one that reflected his focus on biomedical research and his inattention to the social sciences. In its introduction, for example, it stated that "scientific research in earlier times did not commonly involve research on human subjects. It was largely practiced in the physical sciences, and subsequently in anatomy and physiology of animals."[42] This statement can only be true if "scientific research" is taken to exclude centuries of social inquiry.

Beauchamp's draft, like the 1976 draft before it, cited key documents about medical ethics while ignoring the ethical codes of social science. The Nuremberg, Helsinki, and DHEW codes all appeared, along with repeated references to the Hippocratic Oath. Beauchamp did add a behavioral element by citing the American Psychological Association's principle that "the ethical investigator protects participants from physical and mental discomfort, harm, and danger." But he considered psychology representative of all "behavioral research," ignoring the questions raised by the anthropologists and sociologists who had criticized the commission's work up to that date.[43]

Indeed, Beauchamp's use of the code of the American Psychological Association as the basis for judging all "behavioral research" is ironic, given the origins of that code. In formulating both its initial code of 1953 and its research code of 1972, the APA had specifically rejected what one psychologist termed the "armchair approach" of letting a few eminent psychologists be allowed to dictate the ethics of the whole profession. Instead, the APA embraced "empirical and participatory principles" by involving thousands of members in the deliberations.[44] By failing to solicit the opinions of a wide range of anthropologists, sociologists, and other social researchers, Beauchamp rejected the procedures that had formed the APA code he so admired.

In some ways, Beauchamp's April 1977 version was a humbler document than the March 1976 draft it replaced. Gone entirely was the section on "Norms for the Conduct of Research," the calls for IRBs to consider the long-term consequences of research on nonparticipants, and the alarm about snowball sampling. Likewise, the new paper warned that "just as there can be different moral weights attached to these principles, so there can be different *interpretations* of the principles themselves." Potentially, then, the IRBs could allow social scientists to pursue retributive justice rather than the distributive justice Beauchamp imagined.[45]

That flexibility vanished as the drafters continued their work. In September 1977, three commissioners—Jonsen, Lebacqz, and Brady—along with staff director Yesley and consultant Toulmin, met at Jonsen's San Francisco home to revise the paper.[46] Based on these discussions, in December 1977 the staff completed a new draft that firmly reestablished medical ethics as the centerpiece of the evolving report. This third draft deemphasized its reliance on medical and psychological ethics. Unlike the March 1976 version, with its explicit reliance on the Nuremberg Code, the Declaration of Helsinki, the DHEW regulations, and the American Psychological Association's guidelines, this new draft described "respect of persons, beneficence and justice" as "consonant with the major traditions of Western ethical, political and theological thought represented in the pluralistic culture of the United States." Aside from the Nuremberg Code, the medical and psychological codes are mentioned only in a footnote; they are not quoted at all. This draft called the APA the "best known" code "for the conduct of social and behavioral research," blurring any distinction between social and behavioral science.[47] Downplaying the medical and psychological origins of the three principles suited Beauchamp's belief in the normative universality of those principles.

To a degree, the origin of these three "basic principles" was irrelevant, given their vagueness. Almost everyone would endorse "respect for persons, beneficence, and justice" in some form or another. As Beauchamp later noted, these were not even principles at all, but "headings."[48] Commissioner Albert Jonsen agreed, calling the principles "fairly vapid . . . [They] hardly rise above common sense notions."[49]

The meat of the December draft came in a new section, entitled "Applications," that responded to Congress's instructions in the National Research Act of 1974. That law required the report to consider "the role of assessment of risk-benefit criteria . . . the selection of human subjects for participation in biomedical and behavioral research [and] the nature and definition of informed consent in various research settings"; the new draft addressed each of those issues, pairing them to the three principles of the first section.[50] In a framework suggested by staff director Michael Yesley, the principles of respect for persons, beneficence, and justice corresponded to the requirements of informed consent, risks-benefit assessment, and the just selection of subjects, respectively.[51] The December draft described these last three requirements as "more particular guidelines governing research."[52]

What did the term "research" mean in this sentence? By this point, the commission had already adopted the staff's broad definition, covering any "communication or interpersonal contact between investigator and subject" that was designed to lead to generalizable knowledge. The December 1977 draft offered a version of this definition, noting that "by 'generalizable knowledge' is meant theories, principles, or relationships (or the accumulation of data on which they may be based) that can be corroborated by scientific observation and inferences." But the Belmont draft did not define "human subject," and Beauchamp was still thinking about the kinds of medical and psychological research that had concerned NIH officials since 1966 and Congress since 1972. As he explained in 2007, "the guiding applications are actually a biased model, or something like that, a biased biomedical model."[53]

Indeed, though it did not explicitly state that it was targeted at biomedical and psychological research, the Applications section was clearly written with such research in mind. Thus the section on informed consent advised:

> In the consent situation an explanation should generally be given of at least the following dimensions of the research, as they are applicable: the procedures to be followed in the proposed investigation, the purposes of the research, departures from ordinary practice, any attendant discomforts, inconveniences and other risks

> that may reasonably be foreseen, as well as alternative therapies or services available (if any). In addition, subjects should be informed of the right to withdraw at any time without prejudice and that any data collected during the investigation will be kept confidential.[54]

The qualifiers "in general" and "as applicable" do little to disguise the fact that, as Reiss had warned in his Belmont paper, such procedures make a great deal more sense in a drug trial than in an ethnography. In the latter, the researchers themselves would have hardly any of the required information at the start of a project, and concepts like "ordinary practice" and "alternative therapies" would be meaningless outside of a medical study.[55]

Though the draft did not quote directly from the medical codes, the drafters still relied on them as authorities for an "Applications" section on risk-benefit analysis:

> Implicit in previous codes and federal regulations that specify criteria for the appropriateness of research has been a requirement to determine that risks to a subject are outweighed by the sum of both the anticipated benefit to the subject, if any, and the anticipated benefit to society in the form of the importance of the knowledge to be gained from the research.[56]

Here they must have been thinking of the medical codes, for social science codes contained no such suggestion.[57] As sociologist Carl Klockars had written in an essay he sent to the commission, "I am not aware of any such weighing of risks and benefits ever occurring in the history of [his book,] *The Professional Fence*."[58] And Reiss had explicitly argued that in behavioral research, "cost-benefit decision rules in decisions to grant or withhold approval are both troublesome and inapplicable."[59]

Finally, a new section on "Selection of Subjects"—an application of the third concept, justice—seemed to have only medical research in mind. As Jonsen later explained, the very principle of justice "was suggested by the common but invidious practice of burdening the indigent sick with research whose beneficial products flowed to the better-off: Tuskegee was the shameful reminder of that practice."[60] Such concerns are evident in the language of the draft, which worried about people whose "illness or socioeconomic condition render them especially vulnerable." In contrast, "if a relevant property such as a particular kind of disease affecting only [one] class is present, then it may be justified to involve members of that class in research on their disease, and even to involve them more than persons without the disease normally are involved in research." (Later drafts would add even more medical language, with references to "patients" and those dependent on "public health care.")[61] In a nation still recovering from the shock of the Tuskegee revelations, these were crucial considerations for medical research. But their applicability to social scientists was unclear.

Combining Mishkin's broad definition of research with Beauchamp's narrow conception of the same word produced a yawning non sequitur at the center of the report. In effect, the report now argued that because *some* ethical codes required informed consent, risk-benefit analysis, and an equitable selection of subjects, *all* researchers must meet those standards. By December 1977, then, serious problems were emerging in this latest iteration of the Belmont paper and in its interaction with the IRB report. But by this point, the clock was running down, and the commissioners had little time to think about the social sciences as they began debating the draft report.

DEBATING THE REPORT

After leaving the Belmont paper up to the staff and individual commissioners for most of 1977, in early 1978 the commission as a body resumed discussing the paper at length in its monthly meetings. These discussions revealed disagreement about basic components of the report, including its applicability to nonmedical research. As Jonsen commented on a near-final draft in February 1978, "there is nothing in here that tells us why we are about to make a great big step which we have made from the beginning. Namely, why ought the thing that we are calling research be subject to what we call review?"[62] Two commissioners—Brady and Turtle—took a broad view, suggesting that the ethical standards identified in the report "exist in society and govern all human behavior," including but not limited to research on human subjects.[63] But at the next meeting, in March 1978, the commissioners began to realize the problem of applying medical ethics to nonmedical research. Ryan noted that the commission's September 1977 report on research involving children had been "worked out—largely in a biomedical model, if you will, biomedical and behavioral model," and now DHEW officials were unsure if it was supposed to apply to education research. Cooke confessed his own uncertainty: "I think that some things are applicable, and I suspect some are not."[64]

At that meeting, the commissioners made a faint attempt to debate the ethics of social science research. Cooke, Turtle, and Lebacqz suggested that sociological studies of groups, such as Indian tribes, needed to respect those groups as well as the individuals studied. They were challenged by Seldin—one of the medical researchers on the commission—who believed that the Belmont report should be a "document which deals with handling specific human beings from a medical point of view." Seldin was supported by Toulmin, who argued that "the basic idea that we are concerned with [is] the protection of individual research subjects, who, after all, are the people who are exposed to the experimentation," and by Gray, who warned against forcing sociologists to "respect" noxious groups like the Ku Klux Klan. Lebacqz's proposal for "attention to the interests of specific social units, such as the families or tribes," was defeated.[65]

Beyond this, the commission would not be able to resolve important differences. The commission had little time left—its last substantive meeting was scheduled for April—and its staff was departing; even Beauchamp was working less than half time.[66] It fell to Lebacqz to offer a way out: an admission that the commission had not solved every ethical problem involving the study of human beings. She was particularly troubled by the commission's inattention to policy experiments, like Georgia's demand for copayment by Medicaid recipients, noting, "We have not had sufficient time and discussion to sort out the interfaces between trial programs, pilot projects, and their evaluation and such, and research." So she suggested a footnote indicating "that we have not sorted out all of the interfaces between these sorts of programs and what we are talking about here, that that task remains to be done."[67]

In April, a new draft of the Belmont paper appeared with that note, offering the first clear statement that the report was not meant to apply to all kinds of research. The note read as follows:

> The Commission has not examined the problem of distinguishing social science research and experimentation from 'pilot' or 'demonstration' projects in the delivery of health, education and welfare services by administrative agencies. Because the problems related to social experimentation may differ substantially from those of biomedical and behavioral research, the Commission specifically declines to make any policy determination regarding such research at this time. Rather, the Commission believes that the problem ought to be addressed by one of its successor bodies.[68]

This note raised some important questions. First, what were "social science research and experimentation," terms never defined by the commission or appearing in the law creating it? Second, did the reference to "successor bodies" mean that the report was to be a living document—amended every few years in the light of experience and future study—rather than the one-time declaration it became? And third, what were the implications of this exception for the principles overall? If respect for persons, beneficence, and justice did not—or might not—apply to policy experiments, then they could hardly be the "basic principles, among those generally accepted in our cultural tradition" the report claimed they were. Or did the principles still apply, but only if researchers adapted them using a wholly different set of guidelines from those set forth in the "Applications" section of the report? The footnote conceded that different situations require different ethics, a point absent from the rest of the report. But it did not explore the implications of that concession.

The full footnote survived into the July 1978 draft, but by September the first sentence was deleted, thus eliminating the report's only reference to "social science," as well as its clearest admission that it had not thoroughly examined all the topics on which it was making pronouncements.[69] The truncated footnote could only be even more mystifying. Levine later conceded that, thanks to the footnote, "it does not appear that the Commission was particularly attentive to the problems of social research."[70] Likewise, in 1979 philosopher Finbarr O'Connor observed that "one might read [the footnote] to say the National Commission has nothing to say about social research, and this seems indeed to be what it means." Yet he also noted the commission's attention to social research in its IRB report, complicating the picture.[71]

The final Belmont report also defined research differently from the IRB report, describing it as "an activity designed to test [a] hypothesis, permit conclusions to be drawn, and thereby to develop or contribute to generalizable knowledge (expressed, for example, in theories, principles, and statements of relationships)."[72] This was significantly narrower than the IRB report's definition, which appeared to include qualitative research not designed to test a hypothesis.

Indeed, the commission failed to resolve several differences between the Belmont report and the IRB report. In one way, the Belmont paper—with its requirement that benefits outweigh risks—was stricter than the January 1978 draft of the IRB report, which required only an assessment of risks and benefits, not an outweighing of risks by benefits.[73] The difference was nontrivial, given commissioner Albert Jonsen's belief that "we would do this field a service by expunging from it the language of a favorable risk/benefit ratio."[74] Conversely, the Belmont draft was less strict than the IRB draft in that it offered its principles only "to provide an analytical framework, within which discussion can take place" and cautioned that they "cannot be routinely or mechanically applied so as to resolve beyond dispute particularly ethical problems."[75] Yet the IRB draft listed seven determinations that an IRB must make in its review—a rather routine and mechanical application of some of the Belmont concepts.[76] These discrepancies between the IRB and Belmont drafts survived into the final documents, with the IRB recommendations suggesting only that "risks to subjects be reasonable in relation to anticipated benefits," rather than "favorable," and expanding the routine application from an initial seven into the published IRB report's ten steps.[77]

A final decision on the Belmont draft only heightened the uncertainty. When it published the report, the commission also published revised versions of the background papers prepared for the Belmont meeting in February 1976, presenting them as a two-volume appendix to the report itself.[78] Because some these papers, such as Tropp's, mentioned the social sciences, the publication of the papers could be read as evidence that the *Belmont Report* was intended to cover the social sciences. Or they could be seen as rejected alternatives to the report, especially since the report so clearly disregarded the suggestions of Albert Reiss. (In 1979, Reiss published an essay highly critical of the commission's work, especially the *Belmont Report*, as it applied to the social sciences.)[79] By printing these papers without explanation or commentary, the commission left the question of their relevance up to the reader's imagination. Like the ethical principles listed in the *Belmont Report* itself, they had been stripped of context.

Which was it—a report about medical experiments on humans, or a guide to all human interactions? The truth was that the commission never decided whether the ethical principles it was propounding should apply to social research. The footnote was the closest the commission came to admitting that truth, and even that was not very close. To this day, when American universities pledge that their researchers will abide by the ethical principles of the *Belmont Report*, they are making an ambiguous promise. Should social researchers abide by the whole text of the report—with its concerns about informed consent, risk-benefit analysis, and the selection of subjects? Or are they just to abide by the principles themselves—respect for persons, beneficence, and justice—which, as Beauchamp and Jonsen acknowledge, are so broad as to offer little guidance? Or should IRBs and researchers adhere to the footnote, leaving the ethical problems of social research to be debated by future federal commissions? Vaguely worded and hastily finalized, the *Belmont Report* implies, but does not clearly argue, that the ethics of medical experimentation should govern every action taken by all those who would live a moral life.

Thirty years after the drafting of the *Belmont Report*, two of its main authors disagreed about its applicability to nonmedical research. Commissioner Albert Jonsen agreed with Reiss's views that the principles of medical research could not be applied universally. "We really should have made much clearer distinctions between the various activities called research," he explained. Jonsen further noted that "the principles of the medical model are beneficence—be of benefit and do no harm. I simply don't think that that applies to either the intent or the function of most people doing research."[80] Tom Beauchamp, in contrast, insisted that the Belmont principles—including beneficence—were universal. But even he conceded that "they have to be specified in different ways for . . . different contexts. That's a mighty complicated matter as to how that happens."[81] And the commission had made no effort to specify the rules for social research.

It is not surprising that the National Commission paid so little attention to such fields as sociology, anthropology, linguistics, political science, history, and journalism. Aside from a few brief remarks, Congress had not taken testimony about or expressed interest in these disciplines, nor had it required the commission to investigate them. Aside from Jonsen, no commissioner had significant experience as a social scientist, nor had any members of the commission staff done research in social science ethics. Congress had given the commission a punishing list of tasks concerning the ethics of biomedical research, and even with a doubling of its original two-year lifespan, the commission struggled just to complete those. On the other hand, by the end of the IRB hearings in the summer of 1977, if not before, the commission should have understood the significance of its work for social scientists across the United States. Had it shouldered its responsibility to them, it would have taken their views into account when writing the *Belmont Report*.

Given a greater role in the commission, some social scientists (such as Bernard Barber) might have pressed for guidelines quite similar to those ultimately chosen. But had Beauchamp, and the commission, taken seriously such codes as that of the American Sociological Association, and such advice as Albert Reiss's, they might have concluded that much of social science consists of open interaction between consenting adults; that in such cases, informed consent is vastly more important than any speculation about benevolence or distributive justice, and that muckraking social scientists are sometimes ethically obliged to harm those they study. The *Belmont Report* could have left social scientists who studied only competent adults the single charge, stated in the words of the March 1976 draft, "to take all necessary steps to insure that they do not make promises to their research subjects which they may be unable to keep."[82] Decades later, Beauchamp argued that the report did allow for such possibilities: "Nobody says that each of these conditions has to be satisfied in the sense that no condition can ever be overridden by the weight of the other principles."[83] But the IRB report said just that.

A humbler commission could have acknowledged its lack of expertise in the problems of social science research and declined to make recommendations not grounded in careful investigation. Such a course was hinted at in early 1976, in a letter from Frederick Hoffman, the chair of the IRB at Columbia University's College of Physicians and Surgeons. Hoffman was complaining that the survey researchers employed by Bradford Gray were insufficiently versed in the challenges of biomedical research to ask the right questions in that area. To illustrate the problem, he posed a hypothetical situation:

> If the shoe were "on the other foot," so to speak, and I had been asked to participate in a survey of human investigative studies undertaken by social scientists, I would approach this task with considerable trepidation. I do not "speak their language," I do not know which investigative techniques have been demonstrated to be trustworthy and the value of which of them may have been discredited in a convincing fashion. Because of my profound ignorance of this entire area, I could not calculate prospective benefits and risks with confidence; such calculations, to me, are the critical nub of any human investigative proposal.
>
> Were I a participant in this hypothetical survey of social scientists, I would very much want access to a group of experienced and presumably objective social scientists to whom I could direct my questions and with whom I could discuss problems born of my ignorance.[84]

In short, Hoffman argued that without expert help, biomedical researchers were unqualified to judge the ethics of social scientists, just as social scientists were unqualified to judge the ethics of biomedical researchers.

The commissioners did not deliberately try to impose medical ethics on nonmedical fields, but neither did they take care to define the scope of their work or to consider the consequences of their decisions. Instead, throughout its four-year existence, the commission had ignored vigorous efforts by social scientists to make themselves heard. But even as the National Commission submitted its final reports, a remarkable coalition of scholars was assembling to challenge the commission's legitimacy and its recommendations.

CHAPTER FIVE

THE BATTLE FOR SOCIAL SCIENCE

Though the history of IRB review of the social sciences spans more than forty years, two years—1979 and 1980—stand out as the only period in which social scientists played a significant role in shaping federal and university policies toward their research. The skirmishes at Berkeley and Colorado, the angry letters to the National Commission, and the testimony at the 1977 hearings expanded into a larger national movement bringing together hundreds of scholars. These scholars united behind a single proposal and supported it with essays in the scholarly and popular press and with more subtle lobbying in the executive and legislative branches of the federal government. For a while it seemed as though they would get everything they wanted, and when new regulations were issued in January 1981, they included significant concessions. But social scientists failed to get their preferred language encoded in legislation or in regulations, with severe consequences for future researchers.

THE DRAFT REGULATIONS OF 1979

As the National Commission completed its work, social scientists amplified the criticism they had voiced during the IRB hearings of 1977. In February 1978, Murray Wax—one of the commission's most persistent critics—organized a session on federal regulations and the ethics of social research for the annual meeting of the American Association for the Advancement of Science. He would coedit an anthology on the subject in 1979, filled with essays skeptical of the commission's proposals.[1] In August 1978, a month before the release of the commission's IRB report, the *American Sociologist* ran three articles on research ethics, along with a dozen comments on those articles, many of which reiterated complaints made during the commission's work. Bradford Gray defended that work, claiming that "it would . . . be difficult to argue that social and biomedical research should be conducted according to different sets of ethical principles."[2] But two of the commission's own consultants—Albert Reiss and Charles Fried—had made just that argument at the Belmont conference. And when he compared social science to investigative journalism during commission meetings, Gray himself had come close to that position.

In September the National Commission published its IRB report as a standalone document, and in November 1978 it appeared in the *Federal Register*, along with an invitation to submit comments. Many of those comments proved quite critical. Representatives of the National Association of State Universities and Land-Grant Colleges, the Association of American Universities, and the American Council on Education, for example, complained that "IRBs, lacking clear instructions from HEW, lacking the license to use common sense, are increasingly engaged in setting standards for social and behavioral science research that might be wholly appropriate for most biomedical research, but is [*sic*] in no way useful in protecting subjects of research conducted by social scientists."[3] Daniel Patrick Moynihan—a prominent social scientist and a U.S. senator—warned DHEW secretary Joseph Califano that "neither the current regulations nor the report of the Commission shows a proper appreciation of the distinctive nature of *social science* research and of the perverse and needlessly chilling effect that well-meaning controls may have on it."[4]

Meanwhile, DHEW officials began thinking about how to translate the commission's recommendations into new regulations. The task fell to Charles McCarthy, who (as an NIH legislative aide) had helped shape DHEW's response to the Tuskegee scandal as well as the National Research Act. Soon afterwards he had moved to the NIH's Office for Protection from Research Risks, and in 1978 he succeeded Donald Chalkley as director of that office.[5] Knowing full well the rush in which the 1974 regulations were drafted, McCarthy had earlier argued that "when the Commission's Report on IRB's has been published and comments have been received, a general revision of the entire 45 CFR 46 should be undertaken."[6] For most of a year—from September 1978 to August 1979—a committee with representatives of various agencies within the department worked on the task, coming to agreement on a general outline by the end of March 1979.[7]

The March draft adopted the National Commission's definitions of research and human subjects, though with a significant difference. The commission had defined human subject as "a person about whom an investigator (professional or student) conducting scientific research obtains (1) data through intervention or interaction with the person, or (2) identifiable private information."[8] The DHEW draft eliminated the terms "scientific" and "private" in that description.[9] The difference was subtle, but potentially important. As a critic later noted, *scientific research* connoted the testing of a hypothesis, a much narrower scope than all formal interactions. And while requiring review for *private* information might make sense, requiring it for any work uncovering information would seem to demand IRB approval before a social scientist could read the newspaper.[10]

Unlike the commissioners, however, some DHEW officials questioned whether regulations designed for medical experimentation should really be applied to researchers who just watched and talked to adults. In the spring of 1978, Califano asked his assistant Richard Tropp—who had written a paper for the National Commission—to conduct his own investigation of IRB operations. Tropp and his staff talked to IRB members, researchers, and representatives of scholarly associations and, by the fall of 1978, ended up with a picture of IRB operations far darker than the one put forward by the National Commission. "Many investigators have experienced themselves as having been needlessly harassed by IRBs," they reported, "and stories abound of research projects which have been abandoned entirely either because investigators found the regulatory process unreasonably costly in time and energy or because IRBs have imposed requirements . . . which have effectively killed the research." Some IRBs killed research because they didn't understand it; others because it might prove controversial or lead to litigation.[11] Even this was tactful compared to Tropp's real opinion of IRBs: "Most of what they do appears to be unnecessary or even actively harmful."[12]

Tropp's team did think that IRBs could help guard against breaches of confidentiality in social science research, especially when scholars asked about sexual or criminal behavior. To enable this type of review while restraining IRBs, the staff proposed "a graduated system of protection of subjects" that would "remove entirely from IRB oversight those classes of research which involve no more than minimal risk" and only require full IRB review for the most risky. Moreover, the staff sought to "protect investigators against arbitrary and capricious behavior by IRBs" by requiring timely review and forbidding rejection on the grounds of no scientific merit or potential controversy.[13]

These proposals reversed the trajectory of IRB policy within DHEW over the previous twelve years, which had tended to increase IRB authority. In the spring and early summer of 1979, elements of the department debated the issue. The department's Office of the General Counsel argued for the narrowest applicability and the broadest exceptions. This was opposed by the Public Health Service (PHS), which included the Food and Drug Administration (FDA), the Alcohol,

Drug Abuse and Mental Health Administration (ADAMHA), and the National Institutes of Health (NIH), which in turn included McCarthy's OPRR. The debate centered on three questions.

The first question was the degree of risk presented by survey and observational research. Deputy General Counsel Peter B. Hamilton found that "most surveys are innocuous and to require IRBs to look at all such research in order to find the survey that is truly harmful, would be an unwarranted burden on IRBs that would likely distract them from concentrating on more risky research that needs their attention." Hence he suggested that rather than offer expedited review, the regulations explicitly exclude all anonymous surveys and all survey research that did "not deal with sensitive topics, such as sexual behavior, drug or alcohol abuse, illegal conduct, or family planning."[14] More generally, he sought "to remove from IRB review categories of research that only through 'worst case' analysis present any risk to subjects, and are in almost all instances a waste of time for IRBs to review."[15]

In contrast, health officials believed that surveys and observation threatened serious harm in the form of invasion of privacy. In March 1979, ADAMHA's administrator, psychiatrist Gerald Klerman, warned of the potentially "lifelong stigmatizing of individuals as a result of inappropriate or unauthorized disclosure of information about childhood behavior problems or mental illness" and noted that ADAMHA peer review groups had found inadequate protections of confidentiality. He argued that while survey and observational research might merit expedited review, they should not be excluded entirely.[16] By June, the NIH had agreed. A memorandum prepared for the signature of Surgeon General Julius Richmond warned that "unethical invasions of privacy can and have occurred," and that "inadvertent or compulsory disclosure of information collected in such research can have serious consequences for subjects' future employability, family relationships or financial credit." It also suggested that "some surveys can cause psychological distress for subjects."[17]

In making these claims, neither Klerman nor the NIH memo presented examples of harmful or unethical research projects. Nor did they feel they had to. The NIH position hinted that regulation would be required even if the general counsel should show that *all* interview and observational research was innocuous:

> The argument for retaining IRB review of biomedical and behavioral research in these categories is not only based on the need to protect from harm, but to provide an independent social mechanism as well, to ensure that research is ethically

acceptable and that the rights and welfare of subjects will be protected (including, provision for appropriate informed consent procedures, sufficient justification for waiving consent, and establishment of adequate confidentiality procedures).[18]

The second question was whether IRB review was the right tool to protect against the risks of survey and observational research. The Office of General Counsel suggested that "less burdensome requirements might need to be imposed on survey research to provide some assurance against breach of confidentiality," but that "the procedures in Part 46, even as we propose they be revised, are inappropriate for this purpose."[19] The health agencies, in contrast, asserted that no alternative could "provide all of the vital protections which the IRB review process encompasses, including the review of ethical acceptability of the research, adequacy of the informed consent procedures, and procedures for insuring confidentiality of data."[20] It is not clear why the health officials believed that IRBs were effective at these tasks. Just as they had presented no examples of harmful projects, they presented no examples of effective IRB intervention.

Finally, the two sides split over how closely the regulations should follow the specific language of the National Research Act. For example, because the law required only that institutions receiving funds establish "a board . . . to review biomedical and behavioral research," Hamilton suggested requiring only review, not review *and* approval. More significantly, he suggested that institutions receiving DHEW funds should be required to maintain IRBs only for "biomedical or behavioral research involving human subjects" (as specified by the National Research Act), not other categories. As Hamilton noted, "if Congress had wished . . . to cover all human subjects research, rather than just biomedical and behavioral, it could have done so."[21] But the health agencies were reluctant to cede power. The NIH director complained that the general counsel should not have been allowed to draft what the NIH considered "health related regulations" that should have been crafted by the PHS.[22] And while recognizing that the general counsel's version would best "fulfill the literal requirements of the Act," the NIH preferred "a reasonable interpretation of the Act," one that would extend IRB review to projects not funded by DHEW and that was not limited to biomedical and behavioral research.[23]

Despite the debate, both sides agreed that the bulk of social research should be excluded from the requirement of IRB review. Even ADAMHA's Klerman made it clear that he was primarily concerned about protecting "subjects of biomedical and behavioral research in institutions funded by the PHS" and "would be willing to go along with exemptions for research funded by the Office of Education."[24] To this end, his office proposed that the regulations exclude "product and marketing research, historical research, journalistic research, studies on organizations, public opinion polls and management evaluations where the potential for invasion of privacy is absent or minimal," a position adopted by the health agencies as a whole.[25] No one in the department advocated IRB review for surveys, interviews, and observations not directly concerned with health or criminal matters. What was at stake, therefore, was the precise wording of the exemptions for such projects, as well as the applicability of the regulations to surveys, interviews, and observations concerning physical and mental health.

A few months after this debate within DHEW, McCarthy claimed that "after several late-night sessions the Secretary's Office accepted OPRR's position that the regulations could and should apply to behavioral and social science research as well as to biomedical research."[26] The note of triumph in that claim may offer a clue to his motivations—after all, the more research the regulations covered, the more important the job of the director of the Office for Protection from Research Risks. Beyond that, McCarthy's statement was inaccurate in three respects. First, no one had disputed the applicability of the regulations to some behavioral research, such as psychological experiments. Second, neither the OPRR nor any other element within the department had supported the National Commission's recommendation that the regulations should apply to most social science research; Klerman and his allies within the Public Health Service merely wanted them to apply to health research that used techniques shared with the social sciences. And third, the Secretary's Office did not decide the debate in favor of the OPRR, but instead took it to the public—presenting both the general counsel's and the health agencies' proposals as alternatives.

On 14 August 1979 the department published draft regulations that would apply to *all* human subjects research—federally funded or not—"conducted at or supported by any institution receiving funds from the Department for the conduct of research involving human subjects," which in practice meant every research university, or at least every one with a medical school. But it offered two alternative sets of exemptions. Alternative A was the general counsel's version, excluding from review

> research involving solely the use of survey instruments if: (A) results are recorded in such a manner that subjects cannot be reasonably identified, directly or through identifiers linked to the subjects, or (B) the research (although not exempted under clause (A)) does not deal with sensitive topics, such as sexual behavior, drug or alcohol use, illegal conduct, or family planning.

Alternative B, reflecting the view of Klerman and the health agencies, offered exemptions for "survey activities involving solely product or marketing research, journalistic research, historical research, studies of organizations, public opinion polls, or management evaluations, in which the potential for invasion of privacy is absent or minimal."[27]

After months of debate, both sides could agree that the National Commission had exceeded its congressional mandate when it proposed IRB review for every interaction between a researcher and another person. The question on the table was how far to extend, and how best to phrase, the necessary exemptions. As DHEW announced this question, it noted that "these are 'proposed' regulations and public comment on them is encouraged."[28] That public comment was not long in coming.

CRITICS

Among the hundreds of social scientists who protested IRB review in the years 1977 through 1980, the most prominent was Ithiel de Sola Pool, a chaired professor of political science at the Massachusetts Institute of Technology (MIT) and an expert in computers and communications. Pool had great faith in the power of social science research, and he believed that social scientists should take an active role in shaping society, including public policy. "The only hope for humane government in the future is through the extensive use of the social sciences by government," he wrote in 1967.[29] He was very aware of social research's power to change people's lives, and thus was concerned about the ethics of research, particularly outside of one's own country. Reacting to the Project Camelot controversy of the 1960s, he wrote that American scholars doing research abroad did have ethical obligations to their host countries—in particular, to collaborate with foreign scholars, rather than simply taking information and going home. At the same time, however, Pool objected to any fixed rules, whether imposed by governments or by professional associations, believing instead that "sensitive research should be as far as possible the personal responsibility of the person conducting it."[30] He was equally contemptuous of student radicals who complained about his acceptance of Defense Department funds for his work on digital computing.[31] He would accept funds from any government agency, and restrictions from none.

Pool became involved in IRB issues in 1975, when the MIT IRB told a colleague he could not interview Boston antibusing activists who were breaking the law, on the grounds that his interviews might be used against the criminals. Pool was outraged both that his university would block research on so important a topic, and that it would deploy its power against a part-time assistant professor.[32] He joined an MIT task force on the use of human subjects in social science research. Even then, he sought to distinguish experimental work from mere conversation.

> The [Ethics] Committee's particular concerns are studies that cause pain, risk of injury, or invasion of privacy, that impose the exercise of power on subjects, or trick them by misrepresentation. Studies in which the interaction of experimenter and subject involve no manipulation of the subject beyond free communication between them may also sometimes cause embarrassment, or annoyance, disseminate misinformation, or be of questionable character; but such inevitable consequences of free speech do not constitute grounds for jurisdiction by the Committee or any other authoritative body.[33]

He realized that distinctions were difficult, but he held that "some of the most convenient and efficient ways of protecting human subjects would violate the freedoms we are committed to protect."[34]

In particular, Pool believed that IRB review of interview research threatened two values he held dear. The first was freedom of speech. Pool considered social scientists to be engaged in more or less the same type of project as journalists, whose work had long been constitutionally protected against prior restraint. As he later put it, "the ethics of some of the social sciences, like journalism, may require them to strip the concealing shroud from people whose behavior needs to be exposed for the good of others . . . Historians, political scientists, economists, and many sociologists are in that role. Their professional ethics are in no way inferior to that of the service professions; they are simply different."[35] And, he believed, social scientists deserved the same First Amendment protections accorded to journalists.

Second, he believed that university social scientists merited even *more* freedom than journalists, due to the special nature of universities. A journalist might be the employee of a corporation, subject to his employers' wishes. But, as Pool paraphrased I. I. Rabi, the faculty are not employees of a university; they are the university.[36] Pool deplored the idea that a university administrator could sign an assurance with a federal bureaucrat that would be binding on all researchers affiliated with that university. Pool refused on principle to seek IRB clearance for his own research when it involved only interviews with adults. (He dutifully sought approval when seeking federal funding for an experiment involving showing television programs to children, thinking IRB review of such projects "entirely proper.")[37] He understood that yes, were he to simply hand in the forms, MIT's IRB would approve them without fuss.[38] But if tenured professors at the nation's top universities submitted meekly to censorship, what hope was there for graduate students, untenured professors, and scholars of all ranks who challenged conventional wisdom or studied highly controversial topics?[39]

By the time the National Commission's IRB report was published in September 1978, Pool was a hardened foe of IRB review for survey, interview, and observational research. He was outraged not only by the commission's proposals, but also by its conduct. "The Commission had no social scientists on it and nobody who knew anything about the social sciences on it," he later wrote. "It had one very junior and not very wise social scientist [Bradford Gray] on its staff. It proceeded to pontificate about the application of its rules and ethical principles over vast fields that it knew nothing about."[40]

Thanks to his long and active career as a political scientist, Pool knew scholars, journalists, and public officials around the country. In January 1979 he persuaded journalist Anthony Lewis to address the issue in his *New York Times* column.[41] Lamenting the National Commission's recommendations as "Pettifog on the Potomac," Lewis ridiculed such IRB actions as demanding researchers provide in advance all questions to be asked during an interview, or requiring consent for observations of court cases or political demonstrations. DHEW, he argued, "should apply the rules only where they are needed: for biomedical or similar research, not social science work with adults who know their own minds."[42]

Following this success, Pool wrote to his contacts to ask them to join an emergency committee to protest the National Commission's recommendations. Soon he had recruited some of the most prominent social scientists in the nation—among them Charles Lindblom of Yale; Nathan Glazer of Harvard; James Davis of Harvard and the University of Chicago; William Bouwsma of Berkeley; a group of nine social scientists at the Woodrow Wilson Center; and Steven Muller, the president of the Johns Hopkins University.[43] Antonin Scalia—then a law professor at the University of Chicago—wrote to say that as an IRB member himself he was "disturbed by the authority I find myself and my colleagues wielding over the most innocuous sorts of intellectual inquiry."[44] Eventually, Pool's mailing list contained nearly three hundred names.[45]

Given the effort he had put into responding to the National Commission's recommendations, Pool was deeply disappointed by the draft regulations of August 1979. Pool wrote to his Committee of Concern:

> The draft regulations are outrageous. They take some account of protests HEW has received, but in many ways they are far worse than one could have imagined. Exemptions from review are mainly for powerful well-funded research establishments in education, polling, and policy analysis; individual scholars who wish to interview informally without a schedule or to read in a library are not exempt . . .
>
> This is so outrageous and unconstitutional that one may say to oneself, it can't happen here. Yet HEW has heard these points and still persists. If we relax we will find ourselves at the least in a several year struggle to get enacted rules reversed, or at worst discover free speech undercut while no-one was paying attention.[46]

The critics renewed their efforts. In October, J. W. Peltason, the president of the American Council on Education, called a meeting of representatives of learned societies to coordinate a response.[47]

Part of that response involved a flood of comments on the draft regulations. Of the nearly five hundred letters DHEW received in response to its call for comments, about one-quarter objected to any inclusion of social science in the regulations, many in bitter terms. Sociologists from the University of California, Davis, complained that "these proposals display a profound *ignorance* of social science research requirements, techniques and methodology (the very use of the term, 'human subjects,' is inappropriate) and an unbelievably *arrogant disdain* for First Amendment protections of free speech."[48] Sociologist Lauren Seiler enclosed a paper showing that IRBs were far more likely to modify or forbid proposed surveys than to have any knowledge of a survey that had actually harmed someone. He warned that if the regulations were extended to social science, any benefits to research participants would be outweighed by "the losses to academic freedom, civil liberties, the conduct of inquiry, and the bureaucratization of the research process."[49]

Of those who commented on the two lists of proposed exemptions, about 85 percent preferred alternative A, which reflected the general counsel's views and was widely perceived to "provide broader exemptions than alternative B."[50] But this was not to say that these respondents liked alternative A. For example, the executive secretary of the IRB at the University of Illinois at Urbana-Champaign wrote that "all of the research mentioned in alternative A and alternative B should be exempted and if we were forced to make a choice between just those two alternatives, we would endorse alternative A."[51] Seiler demanded all the exemptions in lists A and B plus several more.[52] What the critics really needed was a simple, comprehensive alternative to DHEW's unsatisfactory lists of exemptions.

Such an alternative came from psychologist Edward L. "Pat" Pattullo, the director of the Center for the Behavioral Sciences at Harvard. Pattullo chaired Harvard's IRB and believed that IRB review of nonmedical research was appropriate when subjects were being deceived or otherwise unable to protect their own interests.[53] He was proud of Harvard's system, which left it to researchers to determine which of their projects fell under the IRB's jurisdiction, but which also empowered the IRB "to intervene when it had reason to doubt an investigator's judgment or honesty." This system, he felt, showed respect for both subjects and researchers, "a prime requisite for any effective control."[54] Pattullo sought to persuade, rather than intimidate, researchers, and he ran his IRB with a light touch, rarely requiring written consent forms, never rejecting a project outright, and requiring review only of projects directly funded by DHEW.[55]

Pattullo watched in alarm as the National Commission completed its work in the summer of 1978. In the *American Sociologist*, he argued against "needless, senseless regulation" of the social sciences, complaining that "a sixth-grader with common sense would see that the cost-benefit balance of all that reviewing and approving and informed-consent-documenting is wildly against the rules already on the books, let alone those that are proposed."[56] Later, he warned DHEW that "there is no sieve, consistent with the maintenance of a healthy research enterprise, which will ensure against every possibility of subject abuse," and that overregulation would in fact diminish subject protection by making researchers cynical about the whole review process.[57] When contacted by Pool, he quickly signed on, sending in the first donation to fund Pool's committee.[58]

In late October 1979, Pattullo took part in a conference on "Behavioral and Social Science Research and the Protection of Human Subjects," sponsored by Public Responsibility in Medicine and Research (PRIM&R), a nonprofit group founded in 1974 to promote research ethics. In his presentation, Pattullo ridiculed the draft regulations for their potential to require the review of innocuous activities such as "a eulogist interviewing people for material about the life of James Conant, in order to draw conclusions about the latter's contributions to science and education" and "a professor of English literature interviewing Norman Mailer and friends for material to be included in a critical biography."[59] He also pointed out that when DHEW staffers asked him about his own experience with IRB review, they were conducting research. Had a university professor asked the same questions, DHEW would have required a ponderous review process.

Pattullo acknowledged that talking could hurt. "The fact that a considerable number of social studies have resulted in subjects experiencing boredom, humiliation, self-doubt, and outrage I do not question," he argued. "Further, it would be surprising if there were not others in which breaches of confidentiality, especially, have led to more dire consequences—though I am not aware of any such cases. Nevertheless . . . the possible harm that inheres in most social research is of a kind that we decided long ago we must risk as the necessary price for a free society."[60] He compared social scientists to investigative journalists, evoking the triumph of the press's coverage of Watergate.

Pattullo then offered a formula that, he believed, could distinguish between dramatic interventions—of the sort deployed by some social psychologists—that merited oversight by IRBs and the kinds of research that should proceed without review:

> There should be no requirement for prior review of research utilizing legally competent subjects if that research involves neither deceit, nor intrusion upon the subject's person, nor denial or withholding of accustomed or necessary resources.[61]

Like the DHEW proposals, Pattullo's formula abandoned the 1974 regulations' applicability to undefined "subjects at risk," choosing instead to delineate which activities, rather than what level of risk, required review. Unlike DHEW, Pattullo gave researchers the benefit of the doubt, offering an "opt-in" rather than an "opt-out" scheme. Whereas DHEW's alternatives A and B both proposed to require review for all kinds of interactions *except* those specified, Pattullo exempted all research unless it hit one of four triggers: legal incompetence, deceit, intrusion on the person, or the withholding of resources.[62] The difference was important, because no one could foresee every type of research that would be proposed in the future. Social scientists rallied around Pattullo's proposal. On November 8, twelve scholarly and education associations—including the American Anthropological Association, the American Historical Association, the American Political Science Association, the American Sociological Association, and the Social Science Research Council—endorsed a proposal to insert Pattullo's disclaimer into the regulations, taking the place of either list of exemptions.[63]

Policy makers first reacted with denial. At the PRIM&R conference, former National Commission consultant Robert Levine claimed that thanks to the participation of Gray and consultants like Reiss, the commission had considered the implications of its recommendations for social research.[64] DHEW took a similar line. Charles McCarthy wrote that "all of these objections cited by Dr. Pool had been considered and rejected by the National Commission for the Protection of Human Subjects, and each had been considered and rejected by the Department prior to publication of the [draft regulations]."[65]

As chapters 3 and 4 have shown, Levine and McCarthy were wrong. Indeed, in 2004 both Levine and McCarthy retracted their claims. Levine conceded that "the mainstream of the Commission's deliberations were biomedical . . . They did not spend a lot of time looking at the problems peculiar to social and behavioral science." McCarthy concurred, stating that the commission was "focused on fairly high-powered medical research and unaware that the regulations extended to social science research."[66] And despite his October 1979 dismissal of Pool's complaints, by late November McCarthy had granted one of Pool's top demands: the removal of IRB authority to judge whether a researcher had adopted methods appropriate to his or her question.[67]

Other serious observers realized that the National Commission had failed to explore the ethics of social science. For example, Reiss, whose paper for the Belmont conference was included as an appendix to the *Belmont Report*, distanced himself from that document, complaining that "the language of the biomedical model lends itself to ethical malpractice."[68] In an effort to remedy this oversight, Tom Beauchamp and his colleagues at Georgetown University secured National Science Foundation funding for a two-day conference on the ethics of social science research. Held in September 1979, the conference brought together ethicists and social scientists on a far more equal footing than had been the case with the National Commission. But this conference lacked any official standing and thus could be ignored by the officials in charge of finalizing the regulations. And even if the conference had had some kind of power, its organizers did not seek to achieve the kind of consensus that would have led to specific recommendations for the treatment of social science. Rather, they sought merely to allow various people to air their views, even if they talked past each other.[69]

IRB critics would not get what they wanted by arguing with philosophers. They would have to appeal to people who held political power. And by the end of 1979, Pool had done just that.

THE COMPROMISE OF 1981

In the last months of 1979 and throughout 1980, Pool and Pattullo moved beyond scholarly circles to take their arguments to people with more power to shape federal policy. Some of these efforts were comical. At the June 1980 Democratic National Convention, Pattullo spoke to the platform committee, only to find himself addressing "half a dozen drunks, a handful of nappers, possibly a dozen zombies [who] remained amongst the delegates."[70] Other efforts were more fruitful.

Pool pulled all the strings he could. In a single trip to Washington, D.C., in November 1979, he met not only with McCarthy of the OPRR but also with key congressional staffers, with White House officials, and with staffers at the Brookings Institution and the American Association for the Advancement of Science.[71] The next month, he placed an op-ed piece in the *New York Times*, and soon other periodicals took up the story.[72] The *Nation* editorialized that "in failing to distinguish between medical injections or LSD experiments and survey research or interview procedures customary in the social sciences, the proposed guidelines mark a truly terrifying extension of Federal power in American life," while the *Wall Street Journal* complained that though social research did pose real ethical challenges, "the government has chosen a grossly inappropriate way to regulate these problems."[73]

This pressure had its effect. As early as the end of October 1979, Congressman Edward Markey of Massachusetts had agreed with Pool that "these regulations are overly restrictive for social scientists, and should be limited to biomedical research only," and several other congressmen and senators seemed responsive to Pool's concerns.[74] Congressman David Satterfield, who had served on the conference committee for the original National Research Act of 1974, was particularly outraged by the way DHEW and the National Commission had interpreted that act. He argued that the government should "require IRB review and approval for research only when the Secretary has made a positive finding that the research is of a type which involves potential risk to human subjects," and that no such finding had been made for social science.[75]

That view did not win a majority, but in May Pool's effort with Congress bore fruit, in the form of a House of Representatives bill that would state that IRB requirements did not apply to "research which, as determined under regulations of the Secretary, does not involve risk to human subjects." The Committee on Interstate and Foreign Commerce hoped that "many social science research projects are likely to be exempt from IRB review as not posing a risk to human subjects," and it emphasized that review was only appropriate for research posing "serious psychological or economic risk."[76] Though the language of the bill, authored by Congressman Henry Waxman, required a positive determination that a given type of research was risk free, Pool wrote hopefully that it would "solve most problems" that concerned him.[77]

Meanwhile, Congress had authorized a new President's Commission for the Study of Ethical Problems in Medicine and Biomedical and Behavioral Research, as the successor to the National Commission. The new commission included two members of the old—Albert Jonsen and Patricia King—and its deputy director was Barbara Mishkin, who had also served on the National Commission staff. But the staff of the new commission paid attention to Pool's campaign in the press and in Congress, and it proved more responsive than its predecessor to the concerns of researchers in the social and behavioral sciences.[78]

The President's Commission held a hearing in July 1980 specifically on social and behavioral research, giving Pool and other critics their most official forum yet. Pool, Peltason of the American Council on Education, anthropologist Joan Cassell, and sociologist Russell Dynes reiterated many of the points they had made in print. They were joined by Donna Shalala of the Department of Housing and Urban Development, who warned that "we do not agree that harm has been shown to have occurred in all forms of social science research, and we certainly do not agree that they have been shown to have occurred in such frequency and severity as to require the imposition of a whole new layer of review and restraint, with all its attendant direct and indirect costs."[79]

Speaking on behalf of IRB review of the social sciences were two researchers steeped in biomedical research—medical sociologist Judith Swazey and microbiologist David Kefauver of ADAMHA. Kefauver explained that his agency's studies often combined biomedical, behavioral, and social research, and he favored a single set of rules on the grounds that "the prospect of applying different sets of regulations within the same research project or applying regulations to one piece of the same project and not the other piece is a prospect that is chilling to the most hardened of bureaucrats." While he said this jocularly, Kefauver was dead serious about his wish for absolute uniformity in the regulations' coverage. Any exemption, he warned, would lead to the inference that "it is only the physical integrity of individuals that we are worried about or, even then, only physical integrity if it occurs within the context of biomedical experimentation." Indeed, he continued, "if we deregulate or fail to regulate by topical sensitivity . . . we would have to deregulate biomedical research."[80] This belief that any concession to social scientists would bring down the whole regulatory structure may be the most candid explanation of the health agencies' earlier opposition to Deputy General Counsel Hamilton's proposed exemptions.

The President's Commission was not fully persuaded by either side's arguments. Alexander Capron, the commission's executive director, questioned Pool's charge that IRB review of nonfunded research was unconstitutional. On the other hand, he proved equally skeptical of some of the arguments put forth by the regulations' defenders. Capron all but accused McCarthy of deceiving universities into thinking that DHEW insisted on IRB approval of non-federally funded research, when that was not the department's official interpretation of the National Research Act. And he seemed unhappy with the definition of research crafted by the National Commission—he asked the witnesses if they could write a better definition that would not cover a journalist's book.[81]

In September, President's Commission chairman Morris Abram expressed his concerns in a letter to Patricia Roberts Harris, the secretary of the new Department of Health and Human Services, or HHS. (That spring, DHEW had been split into HHS, which inherited the health and welfare agencies, and the new Department of Education.) Abram warned, "We believe that efforts to protect human subjects are ultimately disserved by the extension of regulatory procedures to ever broader areas. In a word, the full panoply of prior review ought not to apply to activities in which there is no discernable risk to human subjects." He argued that the regulations should only apply to work directly financed by HHS. And he called for exemptions from review for "research involving questionnaires, interviews, or standard educational or psychological tests, in which the agreement of subjects to participate is already an implicit or explicit part of a research process which itself will involve little or no risk."[82]

In that last clause, Abram implicitly rejected Pattullo's argument that even risky conversations with consenting adults should proceed without review. Instead, the commission recommended exemptions for

> research involving solely interview or survey procedures if (a) results are recorded in such a manner that subjects cannot reasonably be identified directly or through identifiers linked to the subjects, or (b) the research does not deal with information which, if confidentiality were breached, could place the subjects at risk of criminal prosecution, civil liability, loss of employment, or other serious adverse consequences, except in settings in which subjects may feel coerced to participate.

It also called for exemptions for surveys and interviews on any topic "if the respondents are elected or appointed public officials or persons running for public office," and for "survey activities involving solely product and marketing research, journalistic research, historical research, studies of organizations, public opinion polls, or management processes," provided "the research presents no risk of harming subjects or of invading their privacy."[83] And it insisted that the regulations apply "only to research with human subjects that is conducted or supported by HHS."[84] Taken together, these provisions exempted many more kinds of research than did either alternative A or B of the 1979 draft regulations. The White House concurred. The directors of the Office of Management and

Budget and the Office of Science and Technology Policy warned that too broad an extension would dilute an IRB's ability to control truly risky research and create "regulatory overkill and unnecessary infringement upon academic freedom."[85]

By the fall of 1980, the draft regulations had been attacked not only by a few individual critics, but also by more than a dozen scholarly organizations, the House of Representatives, the National Commission's official successor, and the White House. Slowly, the pressure reached Charles McCarthy's OPRR. In March 1980, Pattullo reported to Pool that at a recent conference "McCarthy took me aside to assure me that we are no more than a hair's breadth apart as to the wisdom of 'Pattullo's dictum.' *But* it is hard, hard, hard to find just the right words, e.g., half the population is 'legally competent' but everyone knows they're totally hopeless and, e.g., also is an unexplained hypothesis in the P.I.'s mind deceit? The more he talked, the less sanguine I became."[86] In early September, McCarthy met four university lobbyists and indicated that he might accept inserting a version of Pattullo's formula into the expected health bill then being discussed in Congress. But he offered "no firm commitments," and by then the passage of any new health research bill seemed unlikely.[87] (The bill ultimately died in the Senate.)

The final straw came on 4 November 1980, as Ronald Reagan won the presidency. Up to this point, presidential politics had played only a limited role, if any, in the steady expansion of IRB rule across four presidential administrations. But, as McCarthy later recalled about the aftermath of the election, "everybody knew that this was not a time to try to propose a new regulation."[88] Indeed, Secretary Harris decided to promulgate no new regulations during the transition to the new administration.[89] To get around this obstacle, McCarthy began promoting the proposed new rules as a *reduction* in regulations, particularly in nonmedical fields. To make this argument, he had to distort the effects of both the 1974 regulations and their proposed replacements.

First, he painted the status quo in dire terms, claiming that "current regulations extend protections to *all* research involving human subjects in institutions that receive HHS support" and "all behavioral and social science research involving humans is currently regulated by HHS."[90] In a memo prepared for a superior, McCarthy suggested that absent new regulations, HHS would have "to continue to extend coverage of all behavioral and social science research involving human subjects even if the research is essentially risk free."[91] This was doubtful. Though many universities were requiring IRB review of all research, within DHEW the regulations had only been applied to research sponsored by the Public Health Service, and not by other elements in the department.[92] Moreover, the department's own general counsel had, only months before, responded to Pool's *New York Times* op-ed article by stating that "the current policy applies only to research involving human subjects which is conducted or supported by HEW," not unfunded research.[93] True, the department had argued that universities accepting PHS grants (as opposed to DHEW grants in general) were required to review all research, regardless of its funding source. But that policy was based on the department's interpretation of the National Research Act itself, not on anything in the 1974 regulations.[94] New regulations would be neither necessary nor sufficient to ensure its reversal.

Having exaggerated the extent of existing regulations, McCarthy then claimed that the new rules were more lenient: the "proposed new rules would exempt risk-free behavioral and social science research resulting in deregulation of about 80% of research."[95] This figure was almost certainly a fabrication by McCarthy. The record contains no hint of the complex investigation that would have been needed to determine such a statistic, and it tended to vary from one memo to the next. Passing McCarthy's arguments up to Secretary Harris, the director of the NIH promised that the exemptions would cover "all risk-free behavioral and social science research," though he noted that the definition of risk free was debatable.[96] At other times, HHS claimed to exempt between 50 and 80 percent of research. As Alexander Capron of the President's Commission remarked, "fifteen years after this system was first put into place, government ignorance about how it actually operates is staggering . . . The statement that they will exempt from 50 to 80 percent is an indication of how little is known about what is now going on."[97]

McCarthy elaborated his positions in a 30 December 1980 memo for the signature of Julius Richmond, the surgeon general and assistant secretary of health. It was to be sent to outgoing secretary Harris as a statement of the health agencies' reasons for opposing the general counsel's proposal to free up most social research. Some elements of this memo were reasoned, if short on documentation. McCarthy noted that "the Department funds some survey and interview activities in connection with social and mental health services and financial assistance programs, where the privacy of the personal information obtained must be protected—examples include research on health conditions which might result in exclusion of individuals from employment opportunities or insurance coverage, and research on the rehabilitation problems of former criminal offenders, drug users and mental health patients." He then claimed that "the IRB review process is invaluable" in assuring "safeguards necessary to protect sensitive and private information." This was something of an overstatement; neither the

National Commission's IRB report or any other had found IRBs effective, much less invaluable, in promoting the maintenance of confidentiality. True, the National Commission had found that of 2,098 studies, three reported breaches that "had harmed or embarrassed a subject," and that more than 10 percent of these projects lack procedures to protect subjects' confidentiality. But since those figures were not broken down by discipline, it was impossible for McCarthy or other policy makers to know whether those deficient projects were of the sort Pattullo wanted exempted.[98]

Other arguments were more strained, as when McCarthy suggested that Congress sought to regulate social science. Lacking evidence from the National Research Act—the purported basis for the regulations—McCarthy had to reach all the way back to the 1965 letter from Congressman Gallagher and two other members of Congress to Surgeon General Luther Terry, fretting about "the use of personality tests, inventories and questionnaires," especially when "given to young people in our schools." Yet Pattullo's proposed formula applied only to studies of competent adults, so concerns about schoolchildren were a red herring. Moreover, a fifteen-year-old letter from three congressmen, written before the creation of university IRBs, was hardly a good guide to the current views of the House of Representatives, which had just passed a bill demanding the deregulation of risk-free research.

McCarthy's December memo made other, even shakier claims. It argued that the "paucity of documented abuses in this kind of research is due to the success of our policy rather than the absence of need for protections," ignoring the scarcity of documented abuses by social scientists in the years *before* IRB review became common. Finally, McCarthy warned that exempting social science "would place the Department's regulations out of conformity with the Nuremberg Code," even though that code (which required research "based on the results of animal experimentation" before any studies on humans) was clearly designed only for medical experiments.[99]

McCarthy later admitted that some of his tactics had been deceptive. As he explained in 2004, he told the Reagan transition team that the new regulations were less stringent than the old ones. "Of course, they weren't, but they looked like they were because we wrote some exceptions." He pulled a similar ruse with the lame-duck secretary, packaging the new rules as "Diminished Regulations for the Protection of Human Subjects" while trusting "nobody down there in the last weeks of the Harris administration getting ready to leave office would actually read it. So they didn't know what all that was about, but they could read the title."[100]

Why was McCarthy so ruthless? What was his real agenda? Over the course of fifteen months—from October 1979 through December 1980—he had dismissed Pool's objections as old news, then granted some concessions to Pool, then told Pattullo and others that he was close to Pattullo's thinking, then deceived two administrations with a bogus statistic, mislabeled regulations, and nonsensical references to the Nuremberg Code. His statements were so contradictory, it is impossible to know which represented McCarthy's real thinking and which were tricks to mislead his opponents.

Later, in 1983, McCarthy further muddied the waters with a public explanation of his thinking about social science:

> What were the indignities to subjects that we felt needed attention? The Wichita jury bugging case, the tearoom trade research, the decision of a Georgia court concerning Medicaid co-payment experiments—and our own unpleasant memories of Psychology 101 and Sociology 202, when we felt we had better humor our professors (hungry for material they could publish), so we "volunteered" as research subjects rather than risk grade discrimination.[101]

But none of this 1983 account can explain his determined rejection of Pattullo's simple formula. Jury-bugging had been outlawed, and Pattullo's formula allowed for IRB review of the sort of deception used by Laud Humphreys. It also allowed for review when research involved the "denial or withholding of accustomed or necessary resources," such as welfare payments or grades. Since the National Commission had not listed undergraduates as vulnerable populations, and since the HHS version of the regulations empowered the secretary to waive IRB review for "specific research activities" (such as copayment experiments), Pattullo's proposal offered students and welfare recipients better protection than did the HHS proposals. Nor does McCarthy's public statement of 1983 reflect the private documents he produced during the debate itself. One can only guess about his real motives, but a good hint may lie in David Kefauver's warning that to treat each kind of research according to its own needs was "chilling to the most hardened of bureaucrats." By 1980, McCarthy was a hardened bureaucrat, and a hardened NIH bureaucrat at that. If NIH medical officials wanted one thing and university social scientists desired another, he would take care of his own.

McCarthy's stratagems worked, for the regulations did not include the Pattullo formula or anything like it. On 13 January 1981, in the final week of the Carter presidency, Secretary Harris signed the final regulations, which were then promulgated on 26 January, six days into Ronald Reagan's presidency. These new regulations did make significant concessions. First, they eliminated the 1979 draft's requirements that IRBs determine whether each project's "research methods are appropriate to the objectives the research and the field of study."[102] Since this would invite IRBs to meddle with every aspect of project design, Pool considered it "a particularly obnoxious clause" and was glad to see it go.[103] Moreover, contrary to the recommendations of the National Commission, the new regulations offered a limited appeals mechanism. An institution could (but was not required to) establish an appellate body, as long as that body itself met all the regulatory requirements of an IRB.

More significantly, the January 1981 policy greatly reduced the scope of IRB review in two ways. First, the announcement accompanying the new regulations reversed the DHEW/HHS stance, in place since 1974, that the National Research Act required any institution receiving Public Health Service funds to review all human subjects research at that institution, regardless of the funding source for an individual project. Only a year before, the DHEW general counsel had reiterated that position in her letter to the *New York Times*. Now, however, the department announced the opposite: "The HHS General Counsel has advised that there is no clear statutory mandate in the National Research Act to support a requirement for IRB review of other than Public Health Service-funded research." Moreover, the new policy explicitly rejected the National Commission's recommendation that if an institution received any HHS funds, all research at that institution would be subject to IRB review. Instead, the department announced that institutions must only provide a "statement of principles" about the rights and welfare of human subjects, and left IRB review not as a requirement, but as something "strongly recommended." According to the new policy, only research directly funded by HHS would be subject to mandatory IRB review. (Of course, according to the new legal interpretation, there was no statutory authority for even this obligatory review of research supported by an HHS component other than the Public Health Service. But no one seemed willing to quibble, especially now that education research being was being funded by the new Department of Education, not HHS.)

In addition, the regulations offered what HHS termed "broad exemptions of categories of research which normally present little or no risk of harm to subjects." These categories included research on classroom techniques, educational tests where subjects could not be identified, and the reading of publicly available records. The regulations also exempted all

> research involving survey or interview procedures, except where all of the following conditions exist: (i) Responses are recorded in such a manner that the human sub-

> jects can be identified, directly or through identifiers linked to the subjects, (ii) the subject's responses, if they became known outside the research, could reasonably place the subject at risk of criminal or civil liability or be damaging to the subject's financial standing or employability, and (iii) the research deals with sensitive aspects of the subject's own behavior, such as illegal conduct, drug use, sexual behavior, or use of alcohol. All research involving survey or interview procedures is exempt, without exception, when the respondents are elected or appointed public officials or candidates for public office.

Observational research was exempted in an almost identical provision that, mysteriously, did not give blanket permission to observe public officials in public.

This new list of exemptions excused much more research than alternative A of 1979, on which it was based. The 1979 proposal had not exempted interview research of any kind; the 1981 list exempted some interview research—although far less than Pool and Pattullo would have wished. The 1979 proposal would have required IRB review of any survey research in which subjects could be identified and which dealt with sensitive topics. The 1981 regulations only required review of such research if those conditions were met *and* the subject's legal liability, financial standing, or employability were at stake. The 1981 regulations added the freedom to study public officials, a nice sop to Pool, who regularly interviewed such people.

The announcement boasted that taken together, these exemptions would "exclude most social science research projects from the jurisdiction of the regulations," in addition to "nearly all library-based political, literary and historical research, as well as purely observational research in most public contexts, such as behavior on the streets or in crowds."[104] Indeed, since the regulations covered only research that asked about "sensitive aspects," *and* endangered liability or other specific conditions, *and* were funded directly by HHS, observers might reasonably expect that only a handful of projects each year would automatically require review.

Many IRB critics were mollified. Richard Tropp, who had complained of DHEW's failure to consult its own experts in drafting the 1974 regulations, found that the 1981 revision "address[ed] the most critical informed consent issues, and [made] sweeping reductions in the applicability of the regulation to riskless social science research."[105] Richard Louttit of the National Science Foundation claimed that "with minor exceptions, basic research in the following fields is exempt from IRB review: anthropology, economics, education, linguistics, political science, sociology, and much of psychology."[106] The American Sociological Association told its members, "Take heart—most sociological research is now exempt from human subjects regulations."[107] And the *New York Times* reported that the rules appeared "generally to satisfy the complaints of social scientists and historians who had expressed fears that the regulations would unfairly hamper their work."[108]

Pattullo thought that the regulations "still require[d] some protections that are both unnecessary and unwise," but he was delighted that universities would be free to ignore them for research not directly funded by HHS.[109] Believing that McCarthy had been holding back more zealous officials within HHS, Pattullo wrote to McCarthy that the new "rules are sensible, practical, comprehensible and likely to achieve the objective which prompted them" and thanked McCarthy for his "patience, good humor, and quiet common sense."[110]

Ithiel de Sola Pool was less sanguine. He was grateful for the exemptions, he reported, but he still found IRB review inappropriate for some forms of scholarship not covered by the exemptions, such as reading personal correspondence with the permission of the owner, or observing adults' private behavior, again with permission. And while he acknowledged that technically the regulations only applied to HHS grantees, and that few social scientists fit that category, he was concerned that universities would still apply the same policies to all research, regardless of funding, even though HHS no longer required them to do so. He called on scholars to demand that their universities' assurances include a version of Pattullo's formula, and he warned his readers that "our fight is not quite over."[111]

Pool's criticisms of the IRB system could become hyperbolic. In December 1979 he likened Arthur Caplan, a prominent bioethicist, to a "good German" of the early 1930s who watched idly as the Nazis rose to power. Caplan's complacency was shocking, Pool wrote, in the face of "the most vicious attack on academic freedom in America in our lifetime."[112] This strident language no doubt alienated some potential supporters and led opponents, like Levine and McCarthy, to try to dismiss Pool as an angry crank. Yet the coalition that Pool built against the proposed 1979 regulations showed that he was far from alone in his belief that the National Commission had failed to give due consideration to American and scholarly ideals of freedom. He not only succeeded in securing important changes to the regulations, he even forced DHEW/HHS to concede in its 1981 policy that its 1974 legal interpretation of the National Research Act—which it had stubbornly defended for years—was simply wrong.

At the same time, the difficulty Pool faced in gaining these concessions showed what he was up against. One of the nation's most well-connected political scientists had devoted more than two years to his fight. He had won the support of every major social science organization, plus several congressmen and senators. He had presented his position prominently in the *New York Times* and other mainstream journals. And he had done all of this at a time when both major political parties championed deregulation. Yet even then he could not get the Pattullo formula encoded in the regulations, or even extract from McCarthy a straight answer about why that formula was unsatisfactory. The IRB critics had thrown their best punch and landed a blow. But it was not a knockout, and from then on they would keep losing rounds.

CHAPTER SIX

DÉTENTE AND CRACKDOWN

In retrospect, Pattullo and Pool's victory lasted about fifteen years. In that decade and a half, IRBs never quite stopped asserting their jurisdiction over social science and humanities research, but they did not press their cases too hard. Senior scholars in those disciplines, including some who had been active in the debates of the 1970s, stopped worrying about IRBs. Junior scholars—those entering graduate school in the 1980s—often did not even hear of them. Yet these unsupervised researchers managed to avoid the kind of scandal that had surrounded Laud Humphreys.

Meanwhile, however, federal officials quietly reclaimed powers they had ceded. Attracting little attention from social scientists, and soliciting less input, they worked to revise and expand regulations. More significantly, they crafted policies that did not formally make it into the Code of Federal Regulations but had the effect of negating key concessions HHS had made in 1981. Starting around 1995, the Office for Protection from Research Risks came roaring back into social scientists' lives, with more power than ever before.

FEDERAL BACKSLIDING

Federal actions immediately following the promulgation of the 1981 regulations justified Pool's pessimism. While he and Pattullo had secured some freedoms in 1981, they had lost their best chance for a wholesale exclusion of social research from IRBs—as advocated by the Office of General Counsel—or the insertion of a simple formula like Pattullo's into the regulations. The incomplete list of specific exemptions they had achieved soon proved vulnerable to interagency committees, which worked quietly and with almost no public comment. These committees responded to no new research scandals and made no *New York Times* headlines, but they gradually extended IRB review over research that had been exempted in 1981.

The first stage of this backsliding involved the assurances that every institution—usually a university or research hospital—had to file in order to receive PHS research funds. These documents pledged that the institution had established an IRB. Theoretically, each institution was free to draft its own assurance. In practice, the easiest thing to do was to fill in the blanks and sign one of the model assurances that the NIH had distributed since 1966, because they came with the imprimatur of the very agency that could cut off federal funds from an institution that failed to submit a satisfactory assurance.[1] Thus, for individual researchers, the model assurances were almost as powerful as the regulations themselves.

Following the revision of the regulations in 1981, Charles McCarthy of the OPRR circulated a new, twenty-one-page model assurance that violated the spirit of his alleged deregulation. For one thing, despite the January 1981 *Federal Register* pledge to avoid "unnecessary detail" in the requirements for assurances, McCarthy's model was enormously detailed—far more so than previous versions.[2] It indicated that a hypothetical XYZ University submitting the assurance "has chosen to establish an Office of Research Administration to provide a central focus for researchers, IRBs and administrators for processing protocols and communicating other information concerning research involving human subjects." Though the document noted that such an office would not be appropriate for all institutions and was not required by regulations, even the idea that the OPRR was in the business of suggesting how universities might organize their administrations was quite a departure. The model even pledged that each IRB would meet on the third Wednesday of each month. Were a university to submit such an assurance, it would then have to get permission from HHS to move a meeting to, say, the second Wednesday.[3]

The model assurance also diverged from the recommendations of the National Commission and from the newly promulgated regulations of 1981. In some ways, this was good news for social science. In particular, the hypothetical university would establish separate IRBs for biomedical research, for "behavioral and life sciences," and for "social science." By establishing a separate board for social research, the document suggested a way for social scientists to avoid the kind of inexpert IRB review that had caused such problems at the University of Colorado. (Of course, by conceding the distinction between behavioral and social research, the document also called into question whether the regulations applied to the latter at all, since the National Research Act had extended IRB authority only to behavioral research.) And—contrary to the National Commission's recommendations but consistent with the new regulations—it provided for an appeals IRB to oversee the others.

On the other hand, the model assurance gutted a key component of the compromise—the promise that the regulations would apply only to research funded by HHS. In December 1980, McCarthy had made a principled distinction between funded and unfunded research, opposing an exemption for all interview research on the grounds that "risks undertaken by individuals at their own initiative are not comparable to risks undertaken *at the behest of investigators using Department funds*."[4] Yet now, only three months later, McCarthy's model assurance encouraged universities to impose federal standards even on unfunded research.[5]

In correspondence and in journal articles, Pattullo and Pool denounced the double cross. Pattullo complained that the model "includes most of the objectionable features that were excluded from the final regs [*sic*] issued by the Secretary!"[6] He protested to McCarthy that "those who adopt the model will be volunteering to follow a system very similar to that contemplated by the 'proposed' regs of August 1979."[7] Pool noted that "a harried administrator, aware that there is an IRB procedure that is used with HHS-funded research, is likely to respond with a letter saying, 'we will do the same thing with the rest of our research.' Then, voluntarily, social research will be back in the same mess all over again."[8] He hoped that "many social science faculty members and students, as a matter of First Amendment principle, will refuse to submit research on social and political matters to prior censorship by a review board," thus causing university administrations unpleasant controversy.[9]

McCarthy did not try to reconcile the new assurance with his previous distinctions between funded and unfunded research, or with his stated wish to exclude most social science from review. Instead, he merely replied that universities were free to develop alternative assurances. This was rather disingenuous; McCarthy himself noted that more than 90 percent of the institutions that had commented on the model assurance expressed no objection—just as Pool feared.[10] Indeed, by 1985 Pattullo estimated that most universities had signed such assurances.[11] Yet because the model assurances were not legally binding regulations, HHS did not have to submit them to the public comment process that had so empowered dissenters in 1978 and 1979. McCarthy had found a way to cut critics Pool and Pattullo out of the loop.

The second backsliding stage affected the regulations themselves. Despite all the work by the National Commission, DHEW/HHS, and critics outside government, the 1981 regulations had been rushed into print as hastily as the 1974 regulations before them. In May 1980 McCarthy had warned, "These regulations are politically sensitive, and therefore great care in drafting them is needed. To rush the drafting could result in political embarrassment."[12] Yet the drafting had indeed been rushed in a frantic effort to get new regulations signed before Inauguration Day. As McCarthy put it in 2004, "some of the language in those exemptions is so convoluted because it was really a first draft."[13]

With that first draft promulgated as binding regulations, McCarthy set his sights on a new objective: a uniform federal policy on human subjects research. Ironing out the differences in human subjects policy among various federal agencies had been an official goal since at least 1974.[14] But in their haste to have the 1974 and then the 1981 regulations legally declared in some form, officials had postponed the hard work of getting a dozen or more agencies to agree on uniform language, so those regulations applied only to DHEW and to HHS, respectively. With the 1981 regulations in place, officials were ready to try again to write rules across departments.

In December 1981 the President's Commission filed its first biennial report, which recommended that "all federal departments or agencies adopt as a common core" the HHS regulations. In response, the president's science adviser appointed a committee of representatives from more than twenty departments and agencies involved in human subjects research.[15] A second committee, chaired by McCarthy, was appointed in October 1983. Unlike the National Commission or the President's Commission, this new Interagency Human Subjects Coordinating Committee was composed strictly of federal employees. This meant it issued no reports and held no hearings or open meetings that might have alerted social scientists about its actions, nor were its records preserved.[16]

A great deal of the Interagency Committee's work was simply getting a common set of regulations—based on the 1981 HHS code—through the approval process of so many agencies. To the extent that there was serious debate, much of it seems to have concerned parochial requests by participating departments. For example, the Department of Agriculture insisted on an exemption from IRB review "for taste and food quality evaluations," arguing that they posed no risk and needed to be done quickly to correspond to agricultural seasons.[17] The Department of Education called for exemptions of the test, survey, interview, and observation research it sponsored, claiming that the Family Educational Rights and Privacy Act already protected the privacy of subjects of such research.[18] But this only helped educational researchers, not the much broader category of social scientists. The National Endowment for the Humanities—the executive agency most likely to sponsor research in the humanities (broadly enough defined to overlap with some areas in the social sciences)—was not asked to participate, so it could not demand any exemptions for its types of projects. And while the National Science Foundation and the Agency for International Development sponsored social and behavioral research, they sought no departures from the HHS template.[19]

Revising the basic exemptions was not one of the Interagency Committee's main tasks, except to clean up the grammar to make the exceptions easier to understand.[20] But because the compromise of 1981 relied not on Pattullo's straightforward formula but on the admittedly "convoluted" list that McCarthy had crafted, it proved vulnerable to even small changes. The first draft of a Model Policy, completed by an earlier committee in September 1982, was closely modeled on the 1981 HHS regulations, and that draft maintained the broad exemptions for survey and interview research and the nearly identical exemptions for observation research.[21] In the spring of 1984, the Office of Management and Budget suggested combining the observation exemptions with the survey and interview exemptions, saving words and ink. In the process, however, a qualifier was lost. The 1981 regulations exempted research unless it "could reasonably place the subject at risk," while the 1983 proposal eliminated the "reasonably," thus triggering IRB review if the research involved any imaginable risk at all to the subjects' legal status or finances.[22]

When the Model Policy was published in the *Federal Register* in June 1986, "reasonably" was restored, but something else was lost: the 1981 regulations' provision that social research be exempted from IRB review unless it "deals with sensitive aspects of the subject's own behavior, such as illegal conduct, drug use, sexual behavior, or use of alcohol."[23] The announcement offered no explanation for this change. A mere five years after McCarthy had boasted of the "deregulation of about 80% of research," the committee he chaired was proposing to bring some unknown quantity of that research back under regulation, without stating any reason for doing so.

Yet social scientists stayed quiet. While the Interagency Committee kept in touch with biomedical researchers, it made no special effort to contact social scientists, trusting them, it seems, to read the *Federal Register*.[24] Unlike the 1979 proposals, which had suggested a wholesale restructuring of 45 CFR 46, this 1986 proposed change affected just a single subparagraph of the regulations, so it was hard for anyone not as obsessive as Ithiel de Sola Pool to grasp its significance. Pool had died in 1984, and no one had emerged to take his place.

The six comments that the committee later chose to paraphrase in the *Federal Register* called for more regulation, not less. One respondent went so far as to suggest that "if interviews yield identifiable data, regardless of the content, the research should be reviewed by an IRB," a recommendation that could have led to IRB review of every college newspaper story. But the suggestion that stuck was "that the language be broadened to show that harming an individual's reputation in the community was a risk as well as financial standing and employability." Despite the objections of the representative of the Agency for International Development, McCarthy's committee agreed, eliminating the exemption for survey, interview, and observation research if

> (i) information obtained is recorded in such a manner that human subjects can be identified, directly or through identifiers linked to the subjects; and (ii) any disclosure of the human subjects' responses outside the research could reasonably place the subjects at risk of criminal or civil liability or be damaging to the subjects' financial standing, employability, or reputation.[25]

This language woke some critics. Several called for much broader exemptions, with one arguing the exemptions were "written primarily for medical and health research and should not apply to involvement of human subjects for general business interviews or surveys," with others advocating an exemption for business research, and with yet another asking for exemptions for all research that posed only minimal risk. One critic complained that "reputation is a subjective term that is difficult to define operationally" and he "suggested that the wording be changed to limit exceptions to specific risks of 'professional and sociological damage.' "

The Interagency Committee noted all of these complaints in its *Federal Register* announcement and replied simply that it "believes that the exemptions are sufficiently clear so that all types of research, not just biomedical or health research, may be reviewed using the specified criteria. In addition, the Committee has indicated that the [policy] provides for the exemption of certain research including much of the research used by business (e.g., survey research) in which there is little or no risk."[26] And that was that. With the deletion of twenty-eight words about sensitive aspects, and the addition of two words—"or reputation"—federal regulators had retaken much of the ground they had ceded ten years earlier.

The new regulations went into effect on 18 June 1991, not just for the Department of Health and Human Services, but for fifteen other departments and agencies as well, earning the regulations the moniker Common Rule. What had begun in 1966 as a policy specifically for Public Health Service grants was now, twenty-five years later, widespread throughout the federal executive branch. Yet the NIH was still in command. The OPRR (an NIH office) remained in charge of interpreting the regulations and monitoring compliance with them, and few if any other signatories to the rules devoted significant staff time to them.[27] The health agencies had conquered.

IRBs IN THE AGE OF DÉTENTE

Social scientists paid little attention while all of this was happening. The reason, it seems, is that they were being left alone. Reconstructing what IRBs did in the 1980s and early 1990s is not easy. Following the National Commission's survey, which itself learned little about the review of social science projects, there were no large-scale studies of IRB operations until the mid-1990s. And to date, there has never been a thorough survey of IRB review of social science research that includes the perspectives of researchers, reviewers, and research participants.[28] Nor did scholars write about IRBs nearly as frequently as they would after 2000. This silence suggests that IRBs touched most social scientists gently, if at all.

This is not to say that IRBs never reviewed social science research. At some universities, IRB review seems to have been routine for some fields in social science and the humanities. One graduate student in communication, for example, not only submitted a dissertation proposal to an IRB, but was able to compare notes with other students about their IRB experiences. Later, that student was glad to have gone through the process: the IRB provided a university attorney who fought a subpoena for that student's interview notes.[29]

Such IRB jurisdiction may have been exceptional. Anthropologists seem to have had few interactions with IRBs. Russell Bernard's 1988 textbook on research methods in cultural anthropology noted that Stanley Milgram's experiments in social psychology would have trouble getting IRB approval, but it did not suggest that IRB approval was necessary for anthropological fieldwork. On the contrary, Bernard suggested that researchers should heed their own consciences: "Above all, be honest to yourself. Ask yourself: Is this ethical? If the answer to yourself is 'No,' then skip it; find another topic." Rather than submitting such decisions to others, Bernard advised, "be prepared to come to decisions that may not be shared by all your colleagues."[30] This advice remained unchanged in the 1994 edition of the same text.[31] Only in 2003 did Bernard add a paragraph stating that IRBs "review and pass judgment on the ethical issues associated with all research on people." And even this appeared in a new chapter on experimental research, leaving the reader unsure about whether the ethical acceptability of ethnographic research was a matter of individual conscience or one that required committee approval.[32] Carolyn Fluehr-Lobban's 1991 anthology on the ethics of anthropology likewise failed to mention IRBs.[33]

Indeed, many anthropologists likely pursued their work innocent of the idea that IRB regulations might apply to them. Some believed the proclamations of 1981, which had suggested that work was subject to review only if it was directly financed by HHS—or another agency that had adopted HHS rules. In 1991, for example, Fluehr-Lobban noted that academic anthropologists were subject to IRB regulation "if their research funding is derived at all from federal monies."[34] This implied that some academic research was *not* derived from federal monies, and that such research was therefore not subject to review. In addition, some anthropologists believed that student work and qualitative work was exempt. In 1987, for example, a professor at the University of Alabama assigned a graduate student several texts about anthropological ethics before sending her out on an exercise in participant observation. But he did not think that "IRB approval was really necessary in a case involving unfunded and essentially ethnographic exercises." Only after a complaint by another professor—a member of the church congregation being studied—did he ask colleagues if IRB review was necessary for such projects, and even then he got conflicting advice. Eventually, he and his student decided to seek IRB approval, and they got it the same day.[35]

In 1992, Rik Scarce had completed a degree in political science and years of a doctoral program in sociology without learning of IRBs. His first exposure to them came when his university denied him legal support on the grounds that he had not submitted his project to IRB review.[36] Other graduate students completed their doctoral training without ever hearing of an IRB. At the University of Chicago, sociology student Mitchell Duneier observed patrons of a Chicago cafeteria, using research and storytelling techniques that were throwbacks to earlier decades. His adviser, Edward Shils, had long studied questions of ethics and IRBs, but he did not even mention the latter to Duneier, probably believing that the 1981 compromise freed Duneier from such concerns.[37] Around the time Duneier was completing his degree, Sudhir Venkatesh entered the same doctoral program. Venkatesh spent years conducting ethnographic research in Chicago's public housing projects—studying drug dealing, gang violence, and prostitution, and at one point breaching confidences that made his informants vulnerable to extortion. Yet no one told him to seek IRB approval.[38]

Some ethnographers did run afoul of IRBs when they crossed the line between ethnography and medical or psychological research. In 1992, for example, ethnographer Stefan Timmermans began observations in a hospital emergency department, watching attempts at resuscitation. He had little difficulty with his university IRB, but the hospital's IRB berated him for lack of statistical rigor in a project intended to be qualitative. Following that meeting—Timmermans called it "a Goffmanian public degradation ceremony"—the hospital IRB demanded that he submit all papers for its review prior to publication. Not willing to endure what he considered "full-fledged censorship," he stopped his observations. Describing this experience in 1995, Timmermans wrote with dismay that "most ethnographers show a surprisingly submissive attitude toward IRB approval."[39] But it may be that they only seemed submissive because few ethnographers of that period had faced the kind of IRB review that Timmermans did.

At the very least, IRB oversight between 1981 and 1995 failed to spark the kind of published debate that had filled books, journals, and newsletters from 1978 through 1981. A 1992 article about IRBs and anthropology had to reach back to essays published in 1979 and 1980 to document social scientists' feelings.[40] The American Sociological Association's newsletter, *Footnotes*, which had carefully tracked the IRB debate from 1979 to 1981, did not run another item on the question until 2000, suggesting a peaceful interlude of almost two decades.[41]

Political science also remained quiet, judging from the lack of published complaints. In 1985 Robert Cleary surveyed every political science department with a PhD program, along with the IRBs at those universities. He found little conflict. Of the twenty department chairs who characterized their IRB experiences, sixteen rated the experience as positive. Based on this data, Cleary concluded that "the work of IRBs is resulting in heightened attention to the protection of human subjects on the part of political science researchers." But Cleary acknowledged an alternate reading of his results: most political scientists may have simply ignored their IRBs. Two department chairs told Cleary that they refused to waste their time replying to a questionnaire on so unimportant a topic as IRBs. If this attitude explains Cleary's nonresponse rate of 53 percent, then many or most political science chairs did not have enough interaction with IRBs to merit comment. And even among the department chairs who did complete the survey, nearly half reported that their programs had had no experience with IRB review since the start of 1983. Two chairs reported that their faculty ignored the IRB unless they were seeking federal grants, and one wrote that "the IRB situation is a mindless joke insofar as the social sciences are concerned." Of the projects that were submitted, 88 percent of them were cleared without modification. In short, the harmony Cleary perceived between political scientists and IRBs was produced in part by the IRBs' very light hand at the time.

Even then, Cleary's study revealed some tensions. Three department chairs (of the fifty-three who replied) reported "significant problems in meeting IRB requirements." Tellingly, the IRB chairs at those institutions were unaware of the problems. And several political scientists were angry that their universities ignored the federal exemption for research on public officials.[42] Indeed, tensions remained high enough that in 1989, the American Political Science Association's Committee on Professional Ethics, Rights, and Freedoms felt it necessary to advise university IRBs that

> (1) in the assessment of risks and benefits attached to a proposed research project, explicit weight be attached both to the general benefit to freedom of inquiry and to the risk to subsequent research proposals contingent on the restriction of current research;
>
> (2) estimates of risk for human subjects shall be confined to actual subjects as defined by Federal Regulation 45 CFR 46.102 (f) (March 8, 1983);
>
> (3) estimates of risk to human subjects must rise above mere speculation or conjecture.[43]

That same year, the American Sociological Association's new *Code of Ethics* took an ambiguous position on IRBs. The code stated that "study design and information gathering techniques should conform to regulations protecting the rights of human subject [*sic*], irrespective of source of funding, as outlined by the American Association of University Professors (AAUP) in 'Regulations Governing Research on Human Subjects: Academic Freedom and the Institutional Review Board.' "[44] But that very AAUP report suggested that HHS had made a mistake in not adopting broader exemptions for interview and observational researchers, and that a university that imposed HHS standards on interview and observational research not funded by HHS was "highly likely" to violate academic freedom.[45] The ASA's code, then, was less an endorsement of IRB oversight than a recommendation that sociologists comply with a stupid rule.

Even as social science organizations sought to make peace with the federal regulatory system, that system itself was changing, becoming far more strict. As had been the case for more than three decades, policy makers reacted to scandals —or perceived scandals—in medical research, wholly ignoring other fields. Yet once again, social scientists would find themselves carried along by a wave of research policies directed at biomedical researchers. And this time, they would not find a way out.

FEDERAL CRACKDOWN

In late 1993, reporter Eileen Welsome of the *Albuquerque Tribune* ran a series of articles about eighteen Americans who, in the 1940s, had been injected with plutonium as part of a government-sponsored research project on the effects of radiation. Other newspapers picked up the story and found other examples of radiation experiments, including some using prisoners, mentally handicapped teenagers, and newborn babies.[46] Although some such experiments had long been public knowledge, the reporting returned them to prominence, and the radiation experiments became a second Tuskegee scandal. In January 1994, Senator John Glenn called for "a governmentwide review of all testing programs, from drug tests at the Food and Drug Administration to military tests at the Defense Department, to determine if any improper experiments on humans persist to this day."[47] President Bill Clinton created the Advisory Committee on Human Radiation Experiments (ACHRE) to investigate and make recommendations not only on how the federal government should deal with past abuses, but on how to prevent future ones. The next month, he directed the federal executive agencies to review the 1991 regulations and "to cease immediately sponsoring or conducting any experiments involving humans that do not fully comply with the Federal Policy."[48] Nearly two years later, in October 1995, ACHRE issued its final report. The ACHRE report noted that one of the problems with the IRB system as it existed was that IRBs spent too much time reviewing minimal-risk research, leaving them insufficient time to review the truly risky studies. But though it called for "regulatory relief" for IRBs, the radiation report set off a regulatory chain reaction, as one report became many.[49]

Clinton next created the National Bioethics Advisory Commission (NBAC), while Glenn commissioned a General Accounting Office (GAO) study of the effectiveness of federal oversight.[50] The GAO report, issued in March 1996, noted that the review of paperwork alone was no substitute for onsite inspections of IRBs by OPRR officials.[51] In June 1998, the inspector general of the Department of Health and Human Services issued a four-volume report. And in May 1999 NBAC, while not ready with a formal report, wrote the president that it considered federal protections unevenly enforced and inapplicable to privately funded research, leaving Americans vulnerable.[52]

All of these reports cited dubious medical trials, but none found recent examples of the kind of scandals that had prodded both the Tuskegee and human radiation hearings. The GAO, while citing specific cases in which subjects of medical trials had not been adequately informed of the risks, nevertheless concluded that "current oversight activities are working" and suggested only "continued vigilance," not radical changes.[53] HHS's inspector general noted, "We do not claim that there are widespread abuses of human research subjects."[54] And NBAC found that "the current Federal regulations have served to prevent most recurrences of the gross abuses associated with biomedical research in the earlier part of this century."[55] Moreover, the reports generally acknowledged that there was such a thing as too much vigilance. The GAO report in particular found IRBs already struggling under a heavy workload; it stated that an ideal system would "balance two sometimes competing objectives—the need to protect research subjects from avoidable harm and the desire to minimize regulatory burden on research institutions and their individual scientists."[56] NBAC eventually conceded that agencies found "the interpretation and implementation of the Common Rule confusing and/or unnecessarily burdensome."[57]

Nonetheless, everyone wanted to err on the side of more oversight. Even as they lamented burdensome regulations and overworked IRBs, each report made recommendations that, at universities, would translate into longer consent forms, more training, more paperwork, and more inspections by the OPRR. Early in its term, NBAC resolved that "no person in the United States should be enrolled in research without the twin protections of informed consent by an authorized person and independent review of the risks and benefits of the research," while showing no awareness that, taken literally, such a move would require IRB review of much newspaper journalism.[58] Congress also kept up the pressure. In January 1997, Glenn introduced a bill that would extend the Common Rule to all research conducted in the United States, regardless of federal funding, and provide for up to three years' imprisonment for violators.[59] In June 1998, congressmen mocked the OPRR for not knowing how many IRBs existed, for its paltry investigative staff (only one full-time professional), and for the rarity of its onsite inspections and sanctions against institutions.[60]

The OPRR was now under the leadership of Gary Ellis, a biologist who had taken over in 1993, not long after Charles McCarthy's retirement. Ellis took all this criticism—from NBAC, from Congress, and from the president himself—as a signal to expand the sweep of the regulations. As he later recalled, "the atmosphere for biomedical and behavioral research changed dramatically. It was simply not possible for OPRR to ignore research that might be ambiguous, whether it was biomedical and behavioral, or tangential, or truly beyond. It was not possible to ignore anything. The atmosphere for research involving human subjects of any kind was different after the human radiation experiments than before." Ellis had to prepare for all possibilities, including a rewriting of the regulations or even legislation, along the lines proposed by Glenn. He perceived a trend "towards more universal coverage," so he did not think this was the time to give researchers any slack.[61]

Glenn, Clinton, and the press had expressed no alarm about social science, or even any awareness that it fell under the federal regulatory system they were proposing to expand. But along with the general pressure to tighten oversight over research, Ellis felt unease about some social science projects. For example, he happened to attend a presentation by a scholar studying the ways that therapists take notes on what their patients say. As part of the presentation, the scholar shared copies of a therapist's actual notes of a session with a suicidal woman, redacting the name but including enough information about the patient so that she could have been identified by someone who knew her town. Ellis was appalled by the researcher's sloppiness. He also faced controversy over a survey of middle schoolers that asked an open-ended question about what they did with their friends; some reviewers feared that the children might report criminal or sexual activities. And he told NBAC of his concern about a study of adult literacy that was beyond his jurisdiction because the university involved had not pledged to review non-federally funded research.[62] Such projects were never the subject of an official report, and Ellis later recalled them as having been few in number. But faced with such anecdotes, as Ellis termed them, he chose "to err in the conservative direction" and did nothing toward ACHRE's goal of "regulatory relief."[63]

Instead, Ellis expanded IRB review, in part by redefining the meaning of "exempt" in the Common Rule. In March 1979, when the Office of General Counsel had first proposed its list of activities that were to be exempt from IRB review, it had made it explicit that "the regulations, as proposed, do not require any independent approval of a researcher's conclusion that his or her research is within one of the exceptions. In taking this approach for purposes of discussion, we have weighed the possibility of abuse against the administrative burdens that would be involved in introducing another element of review which might not be very different from the review that the exceptions are intended to eliminate."[64] But this interpretation did not make it into the *Federal Register*, which stated only that "the largest portion of social science research will not be subject to IRB review and approval," without making clear whether the IRB, the researcher, or someone else had to approve an exemption.[65] And the model assurance circulated by McCarthy in the summer of 1981 suggested that while researchers and department heads could make a *preliminary* finding that a project was exempt, researchers still needed to write a detailed protocol with "adequate protection of the rights and welfare of prospective research subjects." In the hypothetical XYZ University signing the assurance, only the "Office of Research Administration" could finally declare the project exempt.[66] As a result, in the early 1980s some IRB chairs began telling researchers that they could not declare their own research exempt, but instead had to submit proposals for IRB review just to determine an exemption. In 1983, Richard Louttit of the National Science Foundation, who had helped draft the 1981 regulations, declared such policies "contradictory." He advised that, as far as grant proposals went, a principal investigator and an institutional official were sufficient judges of what was or was not exempt.[67]

Within HHS, officials took a somewhat more open position, leaving it up to individual institutions to decide who was wise enough to interpret the exemptions. In 1988 the assistant secretary for health told a correspondent, "In some institutions, the official who signs the form HHS-596 indicates on the form whether or not a project is considered exempt from required IRB review; in others, the Principal Investigator makes that determination; and in others, every project must be submitted to the IRB chairperson for a determination." All of these choices, he suggested, complied with the regulations.[68]

Ellis, in May 1995, effectively reversed this stance. In *OPRR Reports*, a publication then sent to 12,000 addressees around the world, he now advised "that investigators should not have the authority to make an independent determination that research involving human subjects is exempt and should be cautioned to check with the IRB or other designated authorities concerning the status of proposed research or changes in ongoing research."[69] Ellis did not see the action as a major reversal, but rather the expression of an "accretion of small changes" that were taking place in response to the human radiation scandal.[70] And, technically, this was mere guidance, not a requirement that the OPRR could enforce. Yet universities predictably followed the advice, and by 1998 nearly three-quarters of IRB administrators surveyed stated that IRBs were routinely involved in deciding whether the exemptions would apply. IRB chairs polled at the same time noted that less than half of the protocols eligible for exemption actually received it.[71] Though Ellis later claimed that the "OPRR never quashed any of the exemptions," his actions greatly reduced the chance that they would be applied.[72]

Ellis took similarly expansive approaches to other decisions. He encouraged universities to "begin thinking about outlier departments that they hadn't previously." When IRBs complained that they couldn't handle the increased workload that resulted, he suggested that universities hire more IRB staff.[73] Ellis also advocated wider IRB jurisdiction when responding to individual queries. For example, a political scientist at Texas A&M University complained when the university's IRB demanded the chance to review a project that would use widely available, anonymous data about elections and public opinion—a category clearly exempted by the Common Rule.[74] A university official put the question directly to Ellis, who advised review by the chair or the full board.[75] Ellis later explained, "The only thing that I heard was the institution's calling, asking me for an official reading on whether it's exempt. And it's not ever going to be exempt, if you call me on the phone and ask me that." He conceded that such decisions left more work for the researcher, but they would not expose the OPRR to a charge of letting a dangerous project through. "At some level," he explained, "I've taken care of the human subject. That's the highest priority. The second priority is self-preservation of the bureaucrat. And the third priority is the researcher." In making these choices, Ellis did not feel bound by the precedents set by McCarthy and other regulators. He felt free to read the regulations and university assurances and decide for himself what they meant.[76]

Ellis's interpretations mattered a great deal once he began what one bioethicist called "a totally unprecedented flurry of enforcement activity."[77] Starting in the summer of 1998, the OPRR increased its number of onsite investigations. It also suspended all HHS-funded research at eight major universities and hospitals, including Duke University, the University of Illinois at Chicago, the University of Colorado, Virginia Commonwealth University, and the University of Alabama at Birmingham.[78] Whatever its merits, the enforcement drive redefined the relationship between the federal government and local IRBs. From the original 1966 policies onward, IRB policy had relied on the idea that a group of researchers and lay members gathered by an institution were competent to judge the ethics of a given project. Now the OPRR told institutions that IRBs would only be considered competent if they "periodically receive[d] interactive or didactic training from expert consultants working in the field of human subject protection."[79] Ellis himself later described his tenure as one of "re-regulation." But, like McCarthy before him, Ellis had diminished social scientists' hard-won concessions while avoiding the notice and comment period that would have accompanied a formal regulation. And while Ellis claimed that he merely wanted institutions to take the regulations seriously, he never scolded an institution for violating the regulations' stated intent of excluding most social science research.[80]

Some observers felt the enforcement drive was an overreaction. As a consultant to the National Commission, Robert Levine had championed IRB review. Now he complained that the OPPR was inflicting drastic punishments for what was often a mere failure of paperwork. "If you don't like the way an I.R.B. is keeping minutes," he groused, "you can say so, but you don't need to close an institution to bring about change of this sort." Ellis's crackdown had increased the workload of each member of his IRB at Yale from about eight hours every two weeks to eleven or twelve hours. Likewise, the University of Washington now required IRB members to read 200-page grant applications, rather than just the sections dealing with risks and benefits. "I'm not sure we're getting a whole lot of information from the time we're spending," an administrator there grumbled.[81] The HHS inspector general's office, while applauding the increase in site visits, warned that the new OPRR activity was not producing desired results; few of the reforms it recommended in 1998 had been enacted in the subsequent two years.[82]

Yet no one in power was willing to take the blame for even one bad project getting through, especially after the death, in 1999, of 18-year-old Jesse Gelsinger, a volunteer in a gene therapy study at the University of Pennsylvania. As a Housing and Urban Development assistant secretary in 1980, Donna Shalala had argued that the harms of social science research had not "occurred in such frequency and severity as to require the imposition of a whole new layer of review and restraint, with all its attendant direct and indirect costs."[83] But in 2000, as HHS secretary, she abandoned the idea that regulation should be proportionate to risk, arguing instead that "even one lapse is one too many."[84] She proposed a raft of new requirements: monitoring plans, training regimens, stricter guidelines on informed consent, rules against conflicts of interest, and even more stepped-up enforcement.[85] Most dramatically, in June 2000 Shalala replaced the OPRR with a new Office for Human Research Protections (OHRP) under the secretary of health, rather than within the NIH. Since the move would bring more resources and move the office higher on the organizational chart, some observers thought it would give more teeth to the research oversight effort.[86] Others viewed it as a way to get rid of Gary Ellis, who was seen as too antagonistic to research.[87]

The first director of the OHRP, anesthesiologist Greg Koski, felt that the crackdown had been somewhat counterproductive. "Within the larger research community, and particularly the IRBs themselves," he later wrote, "the suspensions created a crisis of confidence and a climate of fear, often resulting in inappropriately cautious interpretations and practices that have unnecessarily impeded research without enhancing protections for the participants. Such 'reactive hyperprotectionism' does not usefully serve the research community, the participants or the public, and it should be avoided."[88] At the start of his term, he told NBAC that "I believe that the current model is one that is largely confrontational in its foundation. It is a model that is focused primarily on compliance, and I don't believe that it is well suited to meet the challenges we are going to face in the next two decades of research."[89] He later emphasized that "there is a reason why we have within the regulations categories for full review, expedited review, exemptions, and even categories for research that is not human subject research."[90] But while Koski made the OHRP into a much larger enterprise than the OPRR had ever been, he was unable to change its basic culture.[91]

As a result, both during and after Koski's tenure the OHRP kept tightening its leash on hospitals and universities. For example, in March 2002 the OHRP began telling U.S. institutions to pledge that "all of the Institution's human subject activities and all activities of the [IRBs] designated under the Assurance, regardless of funding source, will be guided by the ethical principles in: (a) the *Belmont Report* . . . or (b) other appropriate ethical standards recognized by Federal Departments and Agencies that have adopted the Federal Policy for the Protection of Human Subjects."[92] This was a direct violation of the regulations' promise that an institution could choose any "appropriate existing code, declaration, or statement of ethical principles, or a statement formulated by the institution itself."[93] While not tremendously consequential, it symbolized the OHRP's growing wish to micromanage institutions and its growing disregard for the freedoms encoded in the regulations.

Observing these changes, bioethicist Jonathan Moreno argued that they marked the end of an era. The regulations of 1974 and 1981 had produced an era of "moderate protectionism . . . compromise that combined substantial researcher discretion with rules enforced by a minimal bureaucracy." For twenty years, he argued, this compromise had held, but the magnitude, complexity, and novelty of medical research had crushed it. In its place, federal regulators were building a new regime of "strong protectionism . . . disinclined to rely, to any substantial degree, on the virtue of scientific investigators for purposes of subject protection."[94] Though no legislation had passed, nor had any regulations changed, the actions of the OPRR and the OHRP between 1995 and 2002 had fundamentally altered relations among researchers, institutions, and the federal government. Power had shifted away from researchers and toward federal overseers. Caught in the middle, universities did whatever was necessary to avoid another blow from the federal hammer.

UNIVERSITIES CRACK DOWN

In 1979 Robert Levine, who had done a great deal of work for the National Commission, promised researchers that "each IRB is an agent of its own institution. It is not a branch office of OPRR. It is not a branch office of any other funding or regulatory agency."[95] This was never really true. Since the early 1970s, when

DHEW had quashed Berkeley's affidavit system, local IRBs had been constrained by the threat of federal disapproval. But Levine's statement became even less true as federal enforcers cracked down on university IRBs, effectively turning them into branch offices.

One effect was that IRBs abandoned their hands-off approach to social science. In 1994, Fluehr-Lobban recalled that "the *modus vivendi* for the 1970s and much of the 1980s was that most of behavioral science research fell into the . . . low-risk category and was therefore exempt from federal regulation." But she had heard enough complaints from anthropologists to suggest that the *modus vivendi* was over, and that "even small-scale, individual research projects are subject to institutional review."[96] In a 1996 essay nostalgic for the methods used by University of Chicago sociologists in the decade and a half after World War II, Alan Sica feared that replicating their work would be made impossible by the "demand for legalistic informed consent documents to be 'served' on every subject scrutinized or spoken with by the romantic sociologist" and by the "hawkeyed overseers of informed consent."[97]

By the mid- to late 1990s, some sociologists seem to have taken IRB review as part of the nature of things. Rik Scarce, who had accepted jail rather than divulge the names of his sources, recommended getting IRB approval before taking students on a field trip to a city council meeting, a religious service, or shopping mall.[98] Another sociologist, writing in 1996, reported getting IRB approval for a whole class in qualitative research methods. He also led his undergraduates through the twenty pages of IRB instructions and forms, staged a mock IRB review session, and assigned readings critical of the federal regulations. "I never realized there was so much (red tape) involved in doing anything as simple as this," one student wrote. "It is certainly something you have to keep in mind when deciding on a research project."[99]

The red tape of 1996 was nothing compared to what emerged after the shutdowns of 1998. Following the suspension of federal research funding at several institutions, in February 2000 the *Chronicle of Higher Education* reported that "across the country, university administrators and researchers are worried, even panicked, that the same thing could happen at their institutions, with millions of dollars of research funds from the National Institutes of Health and pharmaceutical companies at stake."[100] The panic was amplified by universities' increasing dependence on NIH funding, which had doubled between 1998 and 2003.[101]

Universities responded to the federal crackdown by imposing local crackdowns of their own. Duke University, for example, reacted to the 1999 suspension of research at its medical center by ramping up oversight of all research everywhere at Duke; once again, social researchers suffered for the sins of medical scientists.[102] The University of Missouri started reviewing journalism projects after a federal audit of its IRB.[103] And IRB administrators at universities not directly affected by the shutdowns tightened their rules to avoid federal penalties. A sociologist on one IRB watched as the board's jurisdiction expanded from federally funded research to unfunded research by faculty and graduate students, and even to coursework.[104] As an anthropologist and IRB member at the University of Chicago explained, "after the hysteria that set in, I.R.B. review has turned into the equivalent of practicing defense against medical malpractice."[105] Koski, having returned from the OHRP to Harvard, put it more bluntly in 2007: "In the 'cover your ass' mentality that has developed over the last decade, we're now in a situation where IRBs do really foolish, stupid things in the name of protecting human subjects but really to cover themselves."[106]

More significantly, following the federal crackdown, universities placed ethical review in the hands of a new class of administrators. In this, IRBs were just one part of a broader trend in universities that shifted power from faculty to professional staff in matters ranging from classroom allocation to business attire. As political scientist William Waugh noted in 2003, "the [university] bureaucracy is increasingly made up of people who have little or no academic experience and do not understand the academic enterprise," causing tensions between administrations and faculties.[107]

When it came to ethical review, the arrival of these staffers marked a shift in the assumptions underlying the IRB system. In the midst of the 1979 debate, bioethicist and IRB supporter Arthur Caplan had argued in the *New York Times* that "given the fact that the bulk of membership of I.R.B.'s is made up of professional academics, drawn from the university community, one might legitimately begin to ask how chilling the effect of such panels is upon free inquiry in the biomedical and social sciences."[108] Of course, IRBs—dominated as they were by biomedical and psychological researchers—had never represented the university community in any comprehensive (or fair) way. And now, in the late 1990s, Caplan's professional academics were giving way to administrators, or, as they began to style themselves, "IRB professionals."[109] At Northwestern University, for example, the crackdown led to an expansion of the human protections staff from two to twenty-six.[110] Beyond numbers, the professionals gained power through presumed expertise in research ethics and regulation.[111] At Northwestern, social scientists offered suggestions for improving the review process, only to be told by administrators that the OHRP would not approve. As observers noted, "IRB panel members, who were less expert [at discerning OHRP intent] than administrators, had no choice but to accept what they were told by administrators."[112] Nor did panel members have much incentive to challenge what they were told. Unlike an IRB administrator—who was making a career of ethics review—a faculty member might serve on an IRB for just two or three years and not wish to spend much time learning the complexities of the system.

The shift in power from scholars to administrators was reinforced by the creation of several national institutions designed to help IRB administrators do their work. Public Responsibility in Medicine and Research (PRIM&R), a nonprofit organization founded in 1974, led the way. In 1999 it created the Certified IRB Professional program, which offered certification (and the right to put the initials CIP after one's name) to candidates who had two years of "relevant IRB experience." This experience could include many types of administrative work, but not experience as an IRB member or as a researcher; it was a program for administrators rather than for university faculty.[113] PRIM&R also worked to credential whole institutions by founding the Association for the Accreditation of Human Research Protection Programs (AAHRPP). Lastly, in 2000 the University of Miami and the Fred Hutchinson Cancer Research Center established the Collaborative Institutional Training Initiative (CITI), a web-based training program for researchers. By October 2007 the initiative's Web site boasted, "The CITI Program is used by over 830 participating institutions and facilities from around the world [and] over 600,000 people have registered and completed a CITI course."[114]

These initiatives had three things in common. First, they shifted power and responsibility away from researchers and IRB members and toward full-time administrators. It would not be Caplan's imagined "professional academics, drawn from the university community" who traveled to expensive PRIM&R conferences to collect continuing education credits for recertification as CIPs—that was the task of an administrator. Likewise, much of the content of the CITI training program was written not by researchers, but by compliance administrators.[115] Second, all of the programs were heavily medical in their orientation. The CIP program, in particular, suggested that candidates bone up on the Nuremburg Code and the Declaration of Helsinki but did not mention any readings specific to the social sciences.[116] Moreover, once certified, CIPs were advised to seek continuing education credits issued by an accrediting body in nursing, pharmaceutical, or medical education.[117] The AAHRPP program was a bit more inclusive, promising standards that would be "clear, specific, and applicable to research across the full range of settings (e.g., university-based biomedical, behavioral and social science research)."[118] Yet in 2008 its board of directors included several physicians but no members of a university social science department.[119] Finally, none of the three initiatives seemed to consider academic freedom important enough to state it as a fundamental principle.

Just as McCarthy's model assurance of 1981 undermined the hard-won gains of the social scientists' revolt, so too did these unofficial but influential initiatives brush away the critiques and the policy achievements of Pool, Pattullo, and their allies from the late 1970s. For example, in 1981 Pool had bragged about eliminating from the regulations the requirement that IRBs determine that "research methods are appropriate to the objectives of the research and the field of study."[120] But by 2007 the AAHRPP was telling universities that universities should ask "does the research use procedures consistent with sound research design?" and "is the research design sound enough to yield the expected knowledge?"[121] One 2003 study guide, used by administrators, informed readers that only IRB chairs or administrators could determine that a project was exempt from review.[122] This claim ran contrary not only to federal guidance, but also to the textbook the study guide was meant to accompany.[123] The textbook itself, in a chapter on survey research, warned that "a survey study that finds Hispanic or Asian men to have a greater propensity for domestic violence may inappropriately create a social stigma that affects the entire Hispanic American or Asian American community."[124] This warning ignored federal regulations, which—thanks to Bradford Gray—stated that "the IRB should not consider possible long-range effects of applying knowledge gained in the research . . . as among those research risks that fall within the purview of its responsibility."[125]

As survey researcher Norman Bradburn testified in 2000, the effect of this unofficial system was to make IRB oversight ever more strict. IRBs were

> becoming more and more conservative [because] there is a kind of network . . . that administrators of IRB's [use to] communicate with one another and they sort of say here is a new problem, how do you handle that, and then everybody sort of responds. And what happens is the most conservative view wins out because people see, oh, gee, they interpret it that way so maybe we better do it too. So over time I have seen things getting more and more restrictive and that [largely] because there seems to be only one remedy, that is you close down the entire institution, and I have seen certainly, and I am sure others at our university, a marked change in the way IRB's have behaved since, you know, Duke and other places [got] closed down.

Bradburn was not opposed in principle to IRB review of social research, but he complained that "in the last four or five years it has become much more burdensome and much tighter and this is a kind of bureaucratic creep."[126]

Though the OHRP did not repeat the bloodbath of 1998, it did keep up a steady stream of "compliance oversight determination letters" warning of one deficiency or another, though rarely identifying actual harms to research participants. By 2005 it had compiled a list of fifty-three problems that had led to institutions receiving warnings. Unsurprisingly, many involved allegations that an IRB had been too lax with its researchers, but never had the OHRP scolded an IRB for unnecessarily hampering research.[127] Universities responded with ever more forms, procedures, and training, leading two observers to contend that "oversight is continuing to promote a culture of red tape rather than a culture of ethics, and meeting the demands of OHRP entails more and more institutional resources being devoted to smaller and smaller gains for human subject protection."[128] And while research suspensions were rare, they did happen. In the summer of 2008, for example, the OHRP began an investigation of research at Indiana University–Bloomington, leading to the transfer of projects to a more medically oriented IRB in Indianapolis. While the university promised that its IRB members and staff would attend training, conferences, and a half-day retreat with a former OHRP official, students and faculty abandoned research projects or modified them to avoid human interaction.[129]

"Don't talk to the humans!" warned the cover story in the September 2000 issue of *Lingua Franca*, a magazine dedicated to intellectual and academic affairs. Reporter Christopher Shea noted the growing trend for university IRBs to assert their jurisdiction over research using surveys, fieldwork, and even individual interviews, in fields ranging from history to music. He described a world in which all the concessions achieved by Pool and Pattullo had become meaningless in the face of new practices and interpretations, and "even exempt research must be reported to an IRB."[130] Twenty years after Pool's "near victory," the compromise of 1981 had vanished. Social scientists would have to start from scratch.

The fifteen years that followed the compromise of 1981 showed what could happen when social scientists gained enough power to shape federal policy. Pool, Pattullo, and their allies had won scholars a period of peace in which the ethics of research in the social sciences were debated and nurtured by social scientists themselves. Some scholars made poor choices, as they did under earlier regimes and would do later. But the deregulation of the social sciences in 1981 did not lead to a pattern of abuse.

Instead, the crackdown of the 1990s (like the regulatory efforts of the 1960s and 1970s) was a reaction to abuses in medical experimentation, both historic (the radiation experiments of the Cold War) and recent (Gelsinger's death and other documented cases). As in the 1960s and 1970s, social scientists found themselves swept along, not because of anything they had done, but because regulators preferred to control whole universities rather than make careful distinctions among disciplines.

What distinguished the 1990s from earlier decades was the size, strength, and stealth of the regulatory machine. From McCarthy's model assurance of 1981 through Ellis's 1995 guidance that investigators could not be trusted to read the exemptions for themselves, regulators had built a system far stricter than that promised in January 1981. But they had done so quietly, with little opportunity for public comment, no outreach to social science organizations, and no bold enforcement actions—like the one at the University of Colorado in 1976—that might have alerted critics. By the time enforcement did come, in 1998 and afterward, regulators presented the system as a fait accompli. And social scientists found themselves divided in their response.

CHAPTER SEVEN

THE SECOND BATTLE FOR SOCIAL SCIENCE

By 1999, IRBs were causing headaches for social scientists nationwide, leading to complaints reminiscent of those of the late 1970s. Some veterans of that struggle reemerged to champion the old cause; anthropologist Murray Wax, now in his late seventies, reiterated complaints he had made twenty years before.[1] But the most successful leaders of the earlier effort had left the field of battle. Ithiel de Sola Pool had died in 1984, and E. L. Pattullo was comfortably retired. While some scholars tried to revive the earlier resistance to IRBs, others counseled accommodation with the system as it had emerged in the 1990s. Lacking the unity of their predecessors, social scientists in the new century found themselves able neither to shape the IRB regime nor to escape it.

COMPROMISE

Part of the reason that individual scholars did not fight the gradual increase of IRB power is that their professional associations were now willing to adopt some of the ethics and review procedures that they had resisted in the 1970s. Anthropology proved particularly willing to embrace IRBs. This would have been a surprise to anyone familiar with the debates of the 1960s, 1970s, and 1980s. As early as 1966, for example, Margaret Mead had spoken out against the application of medical ethics to anthropological fieldwork, and at the IRB hearings of 1977, the American Anthropological Association had asked that its work be excluded from review. In the peaceful 1980s, anthropologists had moved even further away from the medical ethics encoded in the *Belmont Report*. As a result of the poor academic job market, a greater percentage of American anthropologists found themselves not as passive observers, but as active participants in efforts to bring development to other countries or to engage with social problems at home.

That often meant working for government agencies, making it hard for anthropologists to obey the old prohibition against secret research and the commandment that the researcher's first responsibility was to the people studied. In 1984, the American Anthropological Association considered striking those concepts from its code of ethics. Instead, in 1989 the new National Association of Practicing Anthropologists (an affiliate of the AAA) adopted ethical guidelines that sought to balance the interests of the research sponsor against those of the people studied.[2]

Anthropology reversed its course during the 1990s. In 1994, Carolyn Fluehr-Lobban argued that anthropologists should embrace the ideal of informed consent, in part on "moral and humanistic grounds," and in part because they were being subjected to regulations devised for medical and psychological researchers.[3] Enraged by such arguments, Wax replied that "the dogma of informed consent was generated by a group of philosophers, theologians, and physicians, who knew nothing of social or anthropological research and . . . did not see fit to remedy their ignorance."[4] Though Fluehr-Lobban continued to champion the concept of informed consent, she stopped trying to justify it in terms of regulations. Replying to Wax, she wrote, "In the end, informed consent is not about forms or federal guidelines that restrict anthropologists in their freedom to conduct research." Rather, "anthropological research is no different than any other kind of research involving human participants, whether it is social scientific, behavioral, medical, or biological."[5] Fluehr-Lobban won that argument, and in 1998 the AAA's code of ethics adopted informed consent language for the first time.[6]

The American Sociological Association (ASA) also softened the anti-IRB position it had maintained since 1966. In 1997 the association adopted a new Code of Ethics that came much closer to medical and psychological codes than had previous versions. Perhaps this reflected sociologists' growing interest in matters of physical and mental health, but it may also reflect the background of the ASA's executive director, Felice Levine, a social psychologist.[7] The new code explicitly credited the American Psychological Association's *Ethical Principles of Psychologists and Code of Conduct* as a source, and it borrowed language from federal human subjects regulations. For example, it told sociologists to emphasize that participants could withdraw from a study without penalty, as if the researcher were conducting a drug study rather than participant observation.[8] More strikingly, whereas the 1989 code had presented IRB review as at best a necessary evil, the 1997 code presented it as a positive good. Without stating that sociologists needed to consult IRBs for every fieldwork project, it did suggest that "if the best ethical practice is unclear, sociologists [should] consult with institutional review boards or, in the absence of such review processes, with another authoritative body with expertise on the ethics of research," and made eight other similar references. This was quite a departure from the position of the association in 1977, when its executive officer had complained that "requiring a premature formulation of a specific research proposal has been a very important obstacle to qualitative research."[9]

The 1997 code's drafters seem not to have understood the magnitude of the shift they were making. Survey researcher John M. Kennedy, who chaired the ethics committee that produced the code, later remembered no one having brought up the ASA's earlier anti-IRB stance or having paid much attention to the question of IRB review. Rather than reversing policy, he thought the committee was just offering "a benign assistance to help people who didn't have enough resources to understand the ethical implications of their research." He believed—rightly or wrongly—that most sociologists were already seeking IRB approval, "so it may as well be a resource for them."[10] When Kennedy described a draft version in the ASA newsletter, he did not mention the IRB language as one of the "major changes" to previous codes.[11] Only after the ASA membership approved the code did two sociologists complain that "the ASA has totally capitulated. Its code means that whatever the IRB demands becomes its ethics stance as well."[12]

Historians' organizations also initially sought cooperation with newly assertive IRBs and regulators. In the debates of the 1970s, historians had remained mostly on the sidelines, for the simple reason that IRB review of oral history remained a hypothetical danger. For example, at the University of California, Berkeley, the home of the prestigious Regional Oral History Office since 1954, historians sat out the controversy of the early 1970s, since no one seemed to consider them to be doing "human subject experimentation" under IRB jurisdiction.[13] One National Commission memo mentioned oral history as research that might be regulated under some definitions of human research, but no one on the commission advocated for such regulation.[14] When Pool rounded up support for Pattullo's proposed exemption clause, the American Historical Association endorsed the resolution, but it did not take an active role.[15]

Throughout the 1980s and into the 1990s, most oral historians went about their work with no reason to think that they should consult an IRB. Perhaps the first recorded assertion of IRB jurisdiction over oral history came in 1989. The *Oral History Association Newsletter* reported that an unnamed university had issued new guidelines subjecting oral history to review. The association's executive secretary wasn't sure how to respond, in part because he did not know if the case was "a fluke [or] a sign of the future developments in our field."[16]

The next warning sign came in the fall of 1995. In August, Nina de Angeli Walls, a doctoral student at the University of Delaware, submitted a dissertation on the history of the Philadelphia School of Design for Women. Though most of the dissertation was based on written or printed materials—some going back to the 1850s—Walls had interviewed seven alumnae of the school and corresponded with an eighth.[17] And when Walls submitted her dissertation to the Graduate Office, a staffer flagged it as having failed to gain approval from the university's human subjects review board. After some discussion, the university administration agreed that Walls could get her doctorate, but that all other historians would have to submit their current and future interviewing projects for review.[18] Nothing like this had ever happened at Delaware, and it is not clear what had changed. "I cannot speak to past history of similar projects," warned an associate provost. "What I can say is that in the future, such projects would need to be reviewed."[19] He did suggest that oral history would "usually" be found exempt, but that only the IRB could say so.

Delaware's bewildered oral historians alerted the Oral History Association (OHA) and sent distress cries onto an oral history discussion list. Not all oral historians were so alarmed. Michael Gordon of the University of Wisconsin–Milwaukee wrote that "members of our human subjects review board have occasionally made ridiculous requests for revision of my protocols, but overall they have been fair and helpful." He and his students had worked with the board to make their consent forms as complete and clear as possible, and they had never had to compromise their goals or principles to get an interviewing project approved.[20] Margie McLellan reported that during her graduate training at the University of Minnesota, the IRB had helpfully explained legal requirements for reporting child abuse.[21] But some readers reported fears or problems—especially inappropriate demands for anonymity when the purpose of the interview was to document an individual's story.[22] By 1997 leaders of the Oral History Association and the American Historical Association had heard enough of these concerns that they sought advice from the OPRR.[23] Based in part on its recommendations, the AHA advised historians to "be cognizant of and comply with all laws, regulations, and institutional policies applicable to their research activities."[24]

The historians were not convinced that these laws, regulations, and policies were wise or just, so in September 1997 three leading oral historians—Howard Green, Linda Shopes, and Richard Cándida Smith—met with top federal officials, including Gary Ellis, director of the OPRR, to see if they could cut a deal. Smith later reported that though the meeting was "cordial and collegial," it also exposed significant differences between the ethical judgment of the historians and the officials' interpretation of federal regulations. Two concerns were methodological: the historians wanted to be able to secure signed consent *after* an interview (when the narrator would know what he or she had said), rather than before, as is typical in a medical or psychological study. And they wanted IRBs to understand that anonymity is rare in oral history. More significant was the question of harm. The historians explained that research on public officials, members of the Ku Klux Klan, or representatives of the tobacco industry, for example, might well include "an adversarial component." When dealing with major controversies, they argued, scholars "might determine that the need to discover and make public research findings should have priority over their obligation to protect subjects who have provided research materials. They may make a determination that the public's need to know may have greater moral urgency for them than is allowed for in the federal regulations." But OPRR officials would have none of it, and Smith flagged the issue as a potential source of conflict between the ethics of oral history and the regulations.

Despite this disagreement, the two sides did find some common ground. They agreed that the federal rules and the Oral History Association guidelines "were consistent with each other in tenor," and the OPRR invited the historians to inform IRBs around the country about the guidelines. Ellis suggested that if oral history were "commonly conducted" at a university, then the regulations required the institution to include a historian on its IRB. Finally, Ellis told the historians that his office was preparing to revise its list of minimal risk procedures eligible for expedited review, and he suggested that historians recommend that oral history be added to that list.[25]

Though harboring doubts, the historians accepted this suggestion, and in March 1998—on behalf of the Oral History Association, the American Historical Association, the Organization of American Historians, and the American Studies Association—Shopes (then president of the Oral History Association) formally requested that oral history interviews using informed consent procedures and signed legal release forms be made eligible for expedited review. Interviewers who did not use the procedures and forms, they suggested, merited full IRB review.[26] In November, the OPRR officially added oral history to the list of eligible activities.[27] The deal seemed to give a federal imprimatur to existing professional guidelines without imposing an overly burdensome process. As Don Ritchie, a former OHA president, put it, "any oral historian who follows the principles and standards expressed in the OHA's Evaluation Guidelines should have no trouble meeting the concerns of the Office [for] Protection from Research Risks or of their university review boards . . . Maybe the specific inclusion of oral history in these reviews will help make the rest of the historical profession more aware of appropriate ethical standards for interviewing."[28] Not everyone agreed. Florida State historian Neil Jumonville, whose student found himself subject to an IRB investigation, was appalled that the American Historical Association had so casually signed away historians' freedom.[29] But the spirit of resistance was not yet dead. At the end of the century, social scientists from various disciplines tried to rebuild Pool's coalition if 1979.

FAILED CONSENSUS

The leader of the effort to bring social scientists into common cause was Jonathan Knight of the American Association of University Professors (AAUP). A political scientist and staffer on the association's Committee on Academic Freedom and Tenure, Knight had been somewhat involved in the AAUP's 1981 critique of IRB review of social science research. But throughout the 1980s and early 1990s, he had heard few complaints from social scientists. When that changed in the late 1990s, he decided it was time for a new report.[30] In November 1999, Knight gathered representatives of the American Anthropological Association, the American Folklore Society, the American Historical Association, the American Political Science Association, the American Sociological Association, the Oral History Association, and the Organization of American Historians to share notes. At the meeting they decided that each organization should poll its members about their experiences as researchers and IRB members.[31] Each discipline, along with the AAUP itself, published requests for comments in their newsletters and e-mail lists. The survey did not ask any blunt, yes-or-no questions, such as whether the respondents thought that the benefits of IRB review outweighed the burden, and some responses showed mixed feelings about IRBs. Nor did the associations or the AAUP prepare formal tabulations of the results. Still, on balance social scientists seemed pretty unhappy with IRB review.

The historians were angriest. The 142 members who responded to the American Historical Association's call for comments reported inconsistent applications of the regulations, IRB demands that historians destroy recordings or otherwise violate the principles of historical research, and an atmosphere that deterred researchers from even trying to conduct interviews.[32] Political scientists sent fewer responses, but some were equally outraged. One complained that his IRB administrator "generally wanted extreme statements about risks and benefits that were not at all suited to survey instruments." While he thought the process could potentially help sort out issues of confidentiality, he wished it could be streamlined.[33]

Another warned that "many social scientists are afraid of the committee and I believe that many simply do not submit their studies." A third sent a file of correspondence documenting Gary Ellis's demand that the use of widely available data sets required review by an IRB chair, if not the full board.[34]

Some sociologists and anthropologists voiced similar frustrations. One anthropologist reported that his IRB had taken four months to approve a project, then backdated the approval so that it went on record as having taken only one month. A colleague studying the Sri Lankan civil war had been told not to ask children about violence. She joked that she might as well "just 'modify' the project a bit more and write a Sri Lankan cookbook instead."[35] A sociologist blamed delays of up to four months on "either inefficiency or a deliberate neglect for a social science operation like ours."[36] But others were more ambivalent. As one sociologist and IRB member put it, "the necessity to review all projects seems to fly in the face of academic freedom, but I still learn alot [*sic*] on the committee."[37]

Despite the responses showing real anger toward IRB review, the AAUP group's representatives from anthropology, political science, and sociology were reluctant to condemn IRBs. Felice Levine, the executive director of the American Sociological Association, was particularly insistent that the report not condemn IRBs out of hand. She claimed that "some of the respondents to the ASA query were negative . . . but as many were positive . . . and many were constructive both about their experience serving on IRBs and being reviewed by them." Moreover,

> I don't think that mainstream social scientists resist IRBs or think necessarily that there isn't a generic role for IRBs with all human research (note that the ASA Code refers repeatedly to institutional review boards or in the absence of IRBs to similar bodies with ethics expertise) suggest[ing] that ASA and its voting members overwhelming [*sic*] support an independent review of the human subjects issues. I have heard no crafting of an academic freedom argument from the active research community; if it came up in responses, it was surely low in frequency.[38]

Did Levine's position reflected most sociologists' views? The ASA membership had voted to approve the 1997 ethics code, with its language about IRBs. But since the inclusion of this language had not been presented as a major change in ASA policy, and its authors did not consider it as such, the 1997 vote was hardly the endorsement that Levine suggested. And more sociologists' responses to the AAUP's call for comments were opposed to IRBs than supportive.[39] Still, Levine was willing to give IRBs the benefit of the doubt. Robert Hauck, representing the American Political Science Association, concurred with Levine's moderation.

The historians, by contrast, came away from the experience far more skeptical of IRBs, and they were frustrated by Levine's wish for an inoffensive report. "Some of [your] proposed deletions remove the complaints that the surveys of our memberships elicited about the 'danger to freedom of research,'" wrote Ritchie, representing the Oral History Association. "Isn't that the core of this whole endeavor?"[40]

In the end, Knight toned down the AAUP report enough to persuade Levine to allow the American Sociological Association's name to be used, though Knight gave up the original idea of writing a report that would be officially endorsed by all the social science organizations.[41] The final report noted serious problems with IRB review of the social sciences, quoting one survey respondent's complaint that social researchers "feel confused and cynical, distrustful of the IRB and regulatory process because it really does not seem to apply to them. And, unfortunately, there are unsophisticated IRBs that are readily confused, very risk averse, very heavy handed." It derided as "baffling" the requirement IRBs determine the importance of knowledge expected to result from a study and warned that "the uncertainty about whether any particular research project will be considered important in relation to its risks, and the vagueness of such an inquiry, may dampen enthusiasm for challenging traditional habits of thinking, testing new theories, or criticizing social and political institutions."

On the other hand, the AAUP report shied away from condemning IRBs as illegitimate or as inherently unsuited to review social science. "A university's effort to ensure that all researchers comply with its human-subject regulations does not offend academic freedom," the report stated, though there would be an abridgement of academic freedom "if IRBs sought to ban research deemed offensive, as some might insist should happen with respect to research on abortion or on race and intelligence." And it suggested that social scientists cooperate with IRBs, even joining them in large numbers. Both of these positions marked significant retreats from the AAUP's 1981 stance, which had suggested that any restriction on interview or observational research was "highly likely" to violate academic freedom and called for such research to be exempted from review.[42] If IRBs were to "give thoughtful consideration . . . to the practices and ethics of social science research when reviewing projects proposed by social scientists," the new AAUP report promised, "social scientists will treat IRB decisions with the respect they deserve."[43]

The victor in this debate was Levine, who later became a leading voice for trying to work within the IRB system. In subsequent years she gained appointments to official committees, and several associations—including the American Political

Science Association, the American Psychological Association, the American Sociological Association, the Consortium of Social Science Associations, and the Law and Society Association—endorsed her comments on OHRP proposals.[44] While Levine was aware of problems with IRB review, she was still able to imagine a legitimate role for such review of most social science.[45]

Ironically, just as American social scientists began accepting IRB jurisdiction, scholars abroad—especially in other English-speaking countries—began voicing the same concerns that had moved Pool and Pattullo in the 1970s. Some nations imposed little or no ethical review on the social sciences, but U.S. rules moved abroad with U.S. grants, with collaborations between Americans and foreign researchers, and with foreign scholars' wishes to publish in American journals.[46] Most significantly, starting in the 1990s, other nations and their universities had adopted many of the structures and concepts pioneered in the United States, including committee review, expedited review, minimal risk, and versions of the Belmont principles. Though these nations rejected the term "institutional review board" in favor of the more descriptive "research ethics board" (Canada) or "research ethics committee" (United Kingdom), the basic idea was the same.[47]

Other nations' ethics documents were much more expansive than the American regulations. The British *Research Ethics Framework* of 2005 defined research as "any form of disciplined inquiry that aims to contribute to a body of knowledge or theory," and a human participant as any living human being.[48] The 2007 Australian National Statement on Ethical Conduct in Human Research, in turn, observed that a similar British definition of human subjects research "could count poetry, painting and performing arts as research," then failed to offer a definition of human subjects research that clearly excludes those endeavors. It then went on to state that "the conduct of human research often has an impact on the lives of others who are not participants," raising the possibility that a novelist might violate Australia's ethical standards without even talking to anyone.[49]

Social scientists had a role in the preparation of these national documents. As a result, unlike the *Belmont Report*, some of these statements did recognize the critical function of some social research. Canada's *Tri-Council Policy Statement* regarding research ethics boards (REBs), for example, noted that "certain types of research—particularly biographies, artistic criticism or public policy research —may legitimately have a negative effect on organizations or public figures in, for example, politics, the arts or business. Such research does not require the consent of the subject, and the research should not be blocked merely on the grounds of harms-benefits analysis because of the potentially negative nature of the findings."[50] Likewise, Britain's *Research Ethics Framework* stated that "some research . . . poses risks to research subjects in a way that is legitimate . . . Much social science research has a critical role to play in exploring and questioning social, cultural and economic structures and processes (for example relating to patterns of power and social inequality), and institutional dynamics and regimes that disadvantage some social groups over others, intentionally or not. Such research results may have a negative impact on some of the research subjects."[51]

While such statements were cognizant of social scientists' concerns, in these nations' application of ethics review—just as in the United States—such fine distinctions often disappeared. As British scholar Robert Dingwall lamented, "U.K. universities have tended to over-comply in the same way as their U.S. counterparts."[52] Likewise, Kevin Haggerty of the University of Alberta observed that "while the *Tri-Council* guidelines accentuate the need for flexibility, the Boards seem to be concerned with a desire for something akin to formal equality, where like cases are treated alike."[53] Even defenders of the guidelines acknowledged that they could be misinterpreted by individual REBs, and they counseled researchers to quote from the guidelines to emphasize their allowance for departure from the medical model.[54] In 2004, a frustrated panel of Canadian scholars in the social sciences and humanities complained of "the granting agencies' desire to create a regulatory structure to deal with the stereotypical clinical trial" that left "REBs that may lack appropriate breadth of expertise free to impose default assumptions that threaten free inquiry for no ethical gain."[55]

In the early 2000s, a team of scholars observed committees in Australia, Canada, New Zealand, the United Kingdom, the United States, and Canada, and found that while accents and acronyms varied, the process exhibited a "remarkable commonality" from one nation to the next.[56] Other scholars concluded that in all of those countries, "social scientists are angry and frustrated."[57]

EXCLUSION

After the failure of 1997–2000 to build a consensus position against IRB review, one group of scholars—historians—continued alone, and for a moment they seemed to achieve a separate victory. Although in some ways the historians' response to IRBs has resembled that of other interview researchers, their long immunity from IRB interference eventually led them to a position distinct from those disciplines involved in the debates of the 1960s and 1970s. For whereas sociologists and anthropologists might argue that they *shouldn't* be subject to IRB regulations, there was little doubt that they *were*, at least for research not exempted by the 1981/1991 versions of the regulations. That much had been made clear by the fights during and after the work of the National Commission. In contrast, the National Commission never explicitly claimed to cover oral history, folklore, or journalism in its recommendations, nor had the authors of any of the various versions of 45 CFR 46. This gave researchers in those fields some space to argue that not only *shouldn't* they be subject to IRBs, they *weren't* under the regulatory definition of research. This argument is best described as one of exclusion. And while folklorists and journalists joined in this debate, historians were the leaders. Their efforts, spectacular but largely unsuccessful, suggest just how hard it could be for scholars to break federal regulators away from their biomedical assumptions.

The historians' quest for IRB exclusion was led by Linda Shopes and Don Ritchie, the same scholars who had, in the late 1990s, reluctantly accepted the OPRR's recommendation that oral history should be placed under IRB jurisdiction but that oral history should also be made explicitly eligible for expedited review. Both Shopes and Ritchie took part in the AAUP's 1999–2000 meetings of social science organizations, with Shopes representing the American Historical Association and Ritchie the Oral History Association. The angry responses they received from historians as part of that work reinforced their earlier skepticism, and by April 2000 Shopes had begun asserting profound differences between the ethics appropriate to historians and the rules imposed by IRBs.[58] Speaking to the National Bioethics Advisory Commission, she complained that "the biomedical and behaviorist frameworks within which 45 CFR 46 was developed have resulted in IRBs' evaluating oral history projects according to standards and protocols not appropriate for historical research, thereby calling into question the underlying assumption of peer review." IRBs, she reported, asked irrelevant questions about recruitment and consent, or made inappropriate demands for detailed questionnaires, the destruction of tapes, the anonymity of narrators, or the avoidance of sensitive topics. Moreover, she reported a "deeper disjunction between the biomedical model of research on which current human subjects regulations are based and the research that historians and perhaps those in other humanities and social science disciplines engage in." IRBs' determination to avoid recording any attributable information about a person's criminal history, she explained, would make it impossible to research the Civil Rights movement. And while historians should treat narrators with honesty and respect, they also embraced "critical inquiry, inquiry that does challenge, that may be adversarial, that may even expose, as interviews with Klansmen and women and with Nazi collaborators, for example, have done."[59]

Historians' recognition of this disjunction led them to seek escape from IRB jurisdiction. Even as they continued to cooperate with the AAUP effort, they were beginning to probe strategies that would exempt oral history from review while leaving other social research covered. In August 2000, Shopes and another historian met with a staffer to Congresswoman Diana DeGette, who was working on a bill to update federal human subjects protections. Their goal was to get as clear an exclusion as possible for oral history, regardless of its effect on other disciplines.[60] In November, Felice Levine came close to endorsing such a separate path, hinting that as *humanities* disciplines, history and journalism had less use for IRBs than did the social sciences.[61]

The National Bioethics Advisory Commission also came close to endorsing that view in its August 2001 report, *Ethical and Policy Issues in Research Involving Human Participants.* The commission found that "certain types of surveys and interviews are considered research, but they can be well managed to avoid harms without federal oversight, as the risks are few and participants are well situated to decide for themselves whether to participate." While in itself it did not offer a better definition of human subjects research than the one contained in existing regulations, *Ethical and Policy Issues* did suggest that the federal government "should initiate a process in which representatives from various disciplines and professions (e.g., social science, humanities, business, public health, and health services) contribute to the development of the definition and the list of research activities subject to the oversight system."[62]

Historians were not sure a new definition was needed. In December 2000, Michael Carhart—a historian who had tangled with the Rutgers IRB—argued that because history is not a science at all, it was not research as defined by the existing federal regulations. "IRB regulations, I believe, are directed at all 'generalizable' research," he explained, "and the whole point of history is that our conclusions are not generalizable. Historians don't try to predict the future like the sciences—social or natural—do; the whole premise of our research is that we describe events that came together under a unique configuration of circumstances."[63]

Carhart was not alone in arguing that oral history was, by definition, free of IRB jurisdiction. In 1999 a working group of the National Science and Technology Council had argued that research subject to the Common Rule "generally does not include . . . journalism, history, biography, philosophy," and several other types of research.[64] Though the working group had no official policy-making status, one of its members, James Shelton, was able to get his agency, the Agency for International Development, to adopt many of its recommendations, including the exclusion of history from review.[65] And in September 2001, the NIH's human subjects protections chief deemed the NIH's own oral history program outside of IRB jurisdiction on the grounds that oral history was not generalizable.[66]

In early 2002, Carhart elaborated on this idea that history did not meet the regulatory definition of research, and he won the interest of Greg Koski, the head of the OHRP.[67] Shopes, representing the American Historical Association, was a bit doubtful at first, fearing that funding agencies might hesitate to give research grants to a discipline that had stated that it was not conducting "research." Shopes also feared that Koski might change his mind.[68] But just as they had taken their cue from the OPRR's Ellis in proposing expedited review as the solution to oral historians' problems, Shopes and Ritchie followed Koski's lead in seeking to be excluded from the definition of research. At the end of January, Shopes declared to the National Human Research Protections Advisory Commission that historians "do not pursue 'generalizable knowledge' as that term seems to be used in 45 CFR 46."[69]

Over the next year and a half, Shopes and Ritchie sought official endorsement of this position, and in May 2003 they were invited to brief OHRP staff on the methods of oral history and on the reasons why they believed it did not fit the regulatory definition of research.[70] That meeting went well enough that the officials asked Shopes and Ritchie for a concrete proposal.[71] So in August 2003, Shopes and Ritchie drafted a statement proclaiming that "most oral history interviewing projects . . . can be excluded from oversight because they do not involve research as defined by the HHS regulations," and that "it is primarily on the grounds that oral history interviews do not contribute to 'generalizable knowledge' that they are not subject to the requirements of the HHS regulations." On 22 September 2003, Michael Carome, the OHRP's associate director for regulatory affairs, wrote to Shopes and Ritchie, concurring with their proposal, though he qualified their claim—that oral history interviews did not contribute to generalizable knowledge—to the more hesitant statement that "oral history interviews, in general, are not designed to contribute to 'generalizable knowledge.' "[72] When the Oral History Association held a conference the next month in Bethesda, Maryland—the OHRP's home town—an OHRP official dropped by to reiterate the decision. Though a little worried about Carome's addition of the term "in general," Ritchie and Shopes declared victory. They called on oral historians to share the new policy with department chairs, deans, university administrators, and anyone else who might be interested. The newsletters of both the AHA and the OHA offered the same, straightforward headline: "Oral History Excluded from IRB Review."[73]

But university IRB administrators were not ready to cede power. On October 30, just five weeks after Carome's letter to the historians, the Office of Protection of Research Subjects at the University of California, Los Angeles, called him to seek more OHRP guidance on "qualitative research utilizing open-ended interviews, especially activities performed by oral historians and other social scientists." A memorandum of that call, prepared by UCLA but apparently approved by Carome, confirmed that "oral history activities, as described to OHRP by the oral history representatives, in general are designed to create a record of specific historical events and, as such, are not intended to contribute to generalizable knowledge." But the memo then went on to discuss three hypothetical projects—apparently drafted by the UCLA IRB with no input from any historian. The interpretations of these examples made the September agreement between the historians and the OHRP meaningless.[74]

The first example seemed promising enough: "an oral history video recording of interviews with holocaust survivors . . . created for viewing in the Holocaust Museum . . . to create a historical record of specific personal events and experiences" would *not* be considered human subjects research. But the second and third examples, "an open ended interview of surviving Gulf War veterans to document their experiences," and "open ended interviews . . . with surviving Negro League Baseball players," *would* be considered research if "the intent of the archive is to create a repository of information for other investigators to conduct research." In other words, oral history would not be regulated if it was designed to "create a historical record," but it would be subject to review if it was designed to "create a repository of information." The memo was silent on what possible legal, ethical, or methodological differences existed between a historical record and a repository of information. In practice, since by definition oral history projects create repositories of information for the use of other investigators, the last two examples made a hash of the OHRP's earlier pronouncement that oral history, *in general,* was excluded from review.

Beyond these baffling examples, the UCLA memo proposed a new definition of generalizable research. Ignoring previous uses of the term in the National Commission's 1978 IRB report, in the *Belmont Report,* and in the regulations themselves, the memo declared that research was "generalizable" if it "designed to draw conclusions, inform policy, or generalize findings." The idea that an intent to inform policy should trigger review was particularly radical. The National Commission—in its IRB report—had advised against considering a project's policy implications when measuring its ethical acceptability, and that idea had been encoded in the 1981 regulations. Moreover, as Charles Fried had pointed out to the National Commission, courts particularly disliked government restrictions on speech when they affected speech about matters of public interest.[75] Yet Carome was now acceding to a proposal that would specifically target such speech.

Finally, the UCLA memo stated that Carome had agreed that "the August 26, 2003 Policy Statement attached to [Dr. Carome's] September 22, 2003 letter was not drafted by OHRP, does not constitute OHRP guidance, and the characterizations of oral history activities in the third paragraph of the Policy Statement alone do not provide sufficient basis for OHRP's determination that oral history activities in general do not involve research as defined by HHS regulations at 45 CFR part 46." After representatives of two historians' organizations had labored a year and a half to meet the OHRP's request for a definition of oral history acceptable to all, the OHRP's Carome disavowed that definition at the first assertion of authority from an IRB. Furthermore, Carome confirmed the key points of the UCLA memo in an e-mail to another university's compliance officer.[76] These communications—unlike the 22 September letter—did not bear Carome's signature on OHRP letterhead, so it is hard to see how the disavowal was any more official than the original waiver. But the damage was done. By December, a version of the UCLA memo (now attributed directly to Carome) was being handed out at a conference attended by IRB administrators from around the country.[77]

Ritchie and Shopes were outraged. "The UCLA document indeed suggests that all archival research is subject to IRB review—in complete contradiction of the earlier policy statement," Shopes later wrote.[78] They complained to Carome, but he simply refused to admit that he had gone back on his word. "OHRP does not believe that there is a conflict between statements made in our September 22 letter to you and our subsequent statements to the many individuals from the IRB and research community who have contacted OHRP," he assured them.[79] In January 2004 he told them that the September letter was still good, but he refused to issue "formal written guidance on oral history activities," as both IRB administrators and oral historians wished. "Given current office priorities," he continued, "OHRP does not anticipate drafting such guidance anytime soon."[80] Five years later, the OHRP still remained silent on the issue.

Predictably, when faced with ambiguity, IRB administrators decided to err on the side of reviewing too much rather than too little. By the end of 2004, the American Historical Association warned its members that many IRBs favored the UCLA document over the letter signed by Carome himself.[81] In a more systematic survey of university policies posted online, the AHA staff found that on "almost 95 percent of the university web sites, the only guidance a faculty member or student will find is a passing mention of oral history among the research methods subject to 'expedited' review," with no mention made of the hard-won exclusion letter of 2003. The AHA promised to keep lobbying federal officials, but its staffers were losing hope. "Given the limited effect of our past communications and agreements with OHRP," they warned, "our expectations are rather limited."[82] After spending most of a decade trying to work with the OHRP, historians had hit a dead end.

Ironically, just as university IRBs around the country were concluding that oral history was subject to review after all, the OHRP launched its own oral history project. In the summer and fall of 2004, senior OHRP staff, including director Bernard Schwetz, interviewed former members, staff, and consultants of the National Commission, as well as former NIH officials. The interviews were videotaped, and both videos and transcripts were posted on the OHRP Web site.[83] Moreover, clips from the interviews were used in a short film made to commemorate the twenty-fifth anniversary of the *Belmont Report*'s publication in the *Federal Register*. Asked in 2007 whether the project had been approved by an IRB, an OHRP official replied:

> OHRP determined that obtaining oral histories of members and staff of the National Commission did not represent research as defined at 45 CFR 46.102(d) because the activity was not a systematic investigation, nor was it intended to contribute to generalizable knowledge. This oral history activity was designed merely to preserve a set of individuals' recollections; therefore, this activity was not subject to IRB review. The interviewers had no specific training related to this activity; and those interviewed did not sign a consent document.[84]

Whatever the OHRP's claims, the Web site seemed suspiciously like "a repository of information for other investigators to conduct research." Indeed, this book draws on those interviews. Moreover, the OHRP's own video used the interviews to draw conclusions that seemed designed to inform policy. But no IRB dared question the practice. For historians—and scholars in other fields who watched the debacle—the lessons were grim. IRBs, such as UCLA's, could not be trusted to consult expert faculty before suggesting policies that would govern the work of those faculty. The OHRP could not be trusted to honor a deal it had suggested and then confirmed with a senior official's signature on agency letterhead. Nor would the OHRP hold itself to the same rules it imposed on universities.

In short, the strategy of exclusion for oral history (and potentially other fields), based on the regulatory definition of research, had worked for at least one group of researchers: OHRP officials, well insulated from the nonguidance they tossed around. It also worked for a handful of academic historians. In late 2007, for example, Columbia University—the birthplace of academic oral history—announced that oral history projects would be considered generalizable, and thus subject to review, only if they would "support or lead to the development of a hypothesis in a manner that would have predictive value," a very small proportion of such projects.[85] By 2009, Amherst College, the University of Michigan, and the University of Nebraska–Lincoln had adopted similar stances, and the University of Missouri–Kansas City had developed a nuanced policy that gave historians considerable discretion while still presenting the IRB as a resource.[86] But these remained exceptions, and as of 2009, most historians who relied on the OHRP's repeated promises of the early 2000s did so at some risk.

Social scientists had much more to fear from IRBs in the years after 1998 than they had in 1978, when Pool began his crusade. True, the 1970s had produced some high-profile cases of interference with social research at Berkeley, the University of Colorado, and Pool's own MIT. But these were few in number compared to the dozens of cases that began to be reported in the late 1990s and early 2000s. Nonetheless, the response of social scientists was timid compared to the challenge they had put up twenty years earlier.

Part of the reason for this relative quiescence was the gradualism with which regulators expanded IRB jurisdiction in the 1980s and 1990s. Whatever its faults, the National Commission had at least been forthright about its intention to rewrite the rules for human subjects research. Its 1978 IRB report gave critics a target against which to rally. In contrast, the expansions of the 1990s and early 2000s were made via quiet changes to the regulations, reinterpretations by the OPRR, phone calls to individual IRBs, and texts aimed only at university administrators. Newcomers to the system, like John Kennedy of the American Sociological Association's ethics committee, could easily mistake what were actually significant changes as simply minor adjustments to rules long in place.

A second factor was one of personalities. There had been only one Ithiel de Sola Pool—a prominent scholar, connected to powerful people in both government and academia, fiercely committed to academic freedom, and willing to devote some of the last years of his life to challenging IRBs. Whether realistic or hyperbolic, Pool's belief in the gravity of the threat had been crucial to winning concessions from federal regulators. Concerned as Jonathan Knight was, he never considered IRB review of social science a threat to academic freedom on a par with McCarthyism, or with the gradual shift to non-tenure-track hiring.[87]

Instead, the scholar willing to devote the most time to the question was Felice Levine, who sought compromise instead of confrontation.

Not everyone agreed with such a stance. Through the first decade of the twenty-first century, scholars remained split on how to deal with the new demands of IRBs. Some, like Levine, sought to make the best of the IRB regime. Others, following the trail blazed by Pool, tried to resist.

CHAPTER EIGHT

ACCOMMODATION OR RESISTANCE?

The decisions of the OPRR and the OHRP in the years after 1995 sparked a resurgence of IRB interest and interference in the social sciences and humanities. In response, scholars began probing the IRB problem more intensely than ever before. While few argued that the system worked well in all cases, scholars disagreed over whether the problems constituted minor errors to be corrected or fundamental flaws in the very concept of prospective review. The former sought ways to live with IRBs, while the latter sought to escape their jurisdiction—positions I will label *accommodation* and *resistance*. The two options were never entirely incompatible. Just as Pool and Pattullo in the late 1970s opposed IRB review of most social science but accepted its utility for some types of research, some researchers in the early 2000s accepted the basic legitimacy of IRBs while asking for profound changes in the IRB regime. On the other hand, the debates between accommodationists and resisters revealed severe disagreements over the nature of scholarly inquiry in the social sciences and humanities.

HORROR STORIES

The upsurge in IRB activity in the 1990s revived a genre of scholarly literature largely dormant since the 1970s. The sad tales told by scholars in their letters to the National Commission and in their testimony at the IRB hearings of 1977 had led Bradford Gray to use the term "horror story" as early as 1978.[1] During the 1980s and 1990s, IRB horror stories grew rare. After 2000, however, journals in several fields began publishing new complaints.

The stories are often published anonymously, or with names and institutions but without dates, so it is hard to use them to map a precise trend in IRB activity.

It is also impossible to say how representative these accounts are. IRB sympathizers may dismiss them as anomalies. "We hear primarily about colleagues' problems with IRBs because, like any bureaucracy, the best boards can aspire to be is well-oiled, smooth-running, and thus silent," asserted sociologist Laura Stark in 2007. "Bureaucracies, by definition, can be effective but not dazzling; yet this makes it tempting to generalize about all boards based on the provocative stories we hear."[2] But there is just as much reason to believe that the problem is, in fact, worse than suggested by those provocative stories, since the IRB abuses reported in print most likely represent a tiny fraction of the total. Few scholars want to antagonize their deans and provosts by speaking publicly about the misdeeds of IRB members and administrators. Like any potential whistleblowers, they have good reason to keep silent—for fear of reprisals—and many stories only come out years after the events, when scholars have safely moved on to new institutions. When Debbie Dougherty and Michael Kramer sought to collect stories about communication scholars' interactions with IRBs, their own IRB insisted that the call for contributions include the warning that "there is a slight risk that your local IRB could discover your participation in this project and take punitive actions against you for what you have written."[3]

Despite such risks—perceived or real—scholars keep writing about maltreatment at the hands of IRBs, and patterns emerge. In many cases, IRBs restored some of the practices that had made social scientists so angry in the 1970s. The clearest case involves inappropriate demands for written consent. In 2002, for example, an IRB demanded that an anthropologist get signed consent forms before talking to illiterate African children.[4] Another IRB told a political scientist to obtain written informed consent before mailing surveys to people who had already agreed to participate in a study of voting behavior.[5] Yet another insisted that researchers first contact all participants by letter. While this might make sense for medical trials in the United States, such a policy arbitrarily limited the ability of field researchers to talk to people they met in public places.[6]

Other boards called for written forms that might endanger the subjects of research. One board insisted that a researcher get signed consent forms from people who had been deemed criminally mentally ill. When the student's advisors pointed out that the consent forms would pose a risk to the subjects (a problem specifically anticipated by federal regulations), the IRB backed down.[7] Another IRB demanded that a researcher give consent forms to victims of abuse, then warn them not to show the forms to anyone. Why, asked the researcher, require the forms at all? Because those are the rules—or so the IRB believed.[8]

Although federal regulations specifically allowed waivers of written consent, IRBs proved reluctant to apply that provision.[9] Even when researchers succeeded in getting waivers, doing so took precious time and effort.[10]

In instances when researchers did plan written consent forms, they found that IRBs insisted on forms that were less accurate or harder to read.[11] In 2003 a team of Canadian researchers sought permission to survey students at four universities. Two of the universities' ethics committees demanded that the survey be accompanied by what the researchers considered to be "unfriendly and overly legalistic cover letters or consent forms," resulting in far lower rates of participation among students there.[12] In other cases, IRB demands were bizarre or even physically impossible. A departmental ethics committee told one student, planning to conduct participant observation, that if she came across someone who had not explicitly consented to be observed, she would have to turn her face away.[13] Another committee insisted that focus group research be conducted anonymously, then refused to explain to a bewildered researcher how this was possible.[14]

IRBs' demands often indicated a refusal to recognize that not all research uses the methods of medical experimentation. Many boards insisted that researchers spell out in advance every decision they would take and every question that they would ask. For experimental research, this is straightforward enough: the whole idea of an experiment is to control all variables but one. But many IRBs seemed not to understand that qualitative researchers don't work that way.[15] Oral historians were asked, at times, to obscure the true names of their narrators or to destroy their interview recordings and transcripts. Both demands would require these historians to violate their professional ethics, and denying narrators the choice to be quoted by name was just plain rude.[16] Some IRB demands had nothing to do with the protection of research participants. Boards felt free, for example, to critique students' research plans and hypotheses, frustrating instructors who wanted students to learn from their mistakes.[17]

At times it seemed as though IRBs were hunting desperately for possible risks, even if they had to be invented. As Kevin Haggerty put it, ethics boards "generally do not know the empirical likelihood of the potential untoward outcomes that they try to regulate," so they guess. And "given that the members of these boards are bright, motivated, well-intentioned, and highly skilled at dealing with hypothetical scenarios, they seem to have no difficulty envisioning any number of potentialities that should be managed through increasingly onerous regulations."[18] One board killed a student's proposed observation of a strip club on the grounds that the student might observe illegal activity—even though there was no reason to think that would happen.[19] Another board demanded that a scholar wanting to interview scientists first warn them that they could be investigated by government agencies, "exploited by hostile entities," or even kidnapped.[20] A British researcher complained, "I was required to get a [criminal background check] before I could research the food practices of my children and their friends, and had to have my friends sign confidentiality and copyright agreements as I served them a cup of tea and a biscuit in my home."[21]

Boards proved particularly fond of speculating that a given line of questioning could traumatize respondents. One group of researchers was "advised that if we were going to have couples talk about conflict issues in our lab, we would need to have a licensed therapist on hand to conduct any counseling that the couples might need as a result of their interaction." After the researchers produced "dozens of research studies from communication journals that used a similar methodology with no mention of having an on-call therapist," the board let the researchers get away with providing phone numbers for the campus counseling center and a domestic violence hotline number.[22] But such accommodations required that research be conducted locally. Another set of researchers wanted to conduct a national survey online, asking participants to recall the death of a family member. The IRB decided that the study could trigger severe psychological distress. Since the researchers could not offer counseling nationwide, the study died.[23] In another case, an IRB forbade a white PhD student from interviewing black PhD students about ethnicity and career expectations because it might be traumatic for the black students.[24] The idea that talking itself was dangerous mystified experienced researchers like Norman Bradburn, who explained that IRBs were mostly concerned about "would people be offended by asking these questions . . . would it be upsetting to them so all of which are kind of speculative in a way and there is tremendous individual variance." By contrast, he insisted, the real ethical issues were whether the answers to the questions would be kept confidential.[25] Interviewees themselves have laughed at consent forms warning them that talking alone could be distressing.[26]

Once IRBs had decided that conversation is dangerous, they naturally began trying to police the most casual of encounters. In Canada, Haggerty—while serving on University of Alberta's Faculty of Arts, Science, and Law Research Ethics Board—found himself reviewing the request of a student "who, as a means to improve her interview skills, wanted to interview her father about his recent vacation."[27] At the University of Colorado, an IRB administrator told students that if they had learned something from an informal conversation that they later wanted to use in their research, they would have to track down their informant to get written permission.[28] The UCLA IRB scolded an undergraduate for drawing on a surprising conversation he had had at a political fundraiser, while the University of North Carolina at Chapel Hill threatened to withhold a degree from a journalism student who had requested printed material by phone.[29]

Perhaps the most serious clashes are those—as predicted by Ithiel de Sola Pool—in which IRBs sought to suppress unpopular ideas. At Florida State University, the IRB application bluntly asked, "Is the research area controversial and is there a possibility your project will generate public concern?"[30] At Simon Fraser University,

> although one of the two community REB [research ethics board] members was a former politician who had engaged in a widely reported disagreement with one of the applicants over a certain aspect of criminal justice policy that she had instigated, that member nevertheless participated in the evaluation of the proposal to conduct research that might produce results that would discredit that policy, despite the researcher objecting that the REB member was in a conflict. The REB then held up the research for eight months without providing any specific reasons as required by policy.[31]

Brigham Young University blocked a study of gay Mormons, while two public universities in California blocked studies on affirmative action and Indian gambling—both particularly sensitive topics in that state.[32]

Researchers who might learn details of criminal behavior had particular trouble. At one university, a scholar wanted to ask parents how they disciplined their children without reporting any abuse to the authorities. Unable to find a way to let the research proceed, the IRB referred it to the university lawyers, where it languished for months until the researcher gave up and withdrew her proposal.[33] Similarly, policy scholar Mark Kleiman was limited in his study of California's probation system. "After considerable delay because one of the 'community members' of the IRB hated the criminal justice system and decided to express that hatred by blocking research about it," Kleiman wrote, "I finally got permission to interview probation officers and judges." But the IRB refused him permission to interview probationers, since it couldn't decide how he might approach them without coercion or risk of exposure. In the name of protecting the probationers, the board denied them the chance to record their views of the system that governed their lives.[34]

Researchers also faced trouble when they wanted to study any kind of minority group. One Northwestern University professor wrote that when reviewing potential research by students, the IRB

> raises its eyebrow at [studies of children and] more nebulous categories, such as gays and lesbians, or poor workers. It basically makes me worried about [studying] any population that is not adult, well-adjusted, crime-free, middle class, heterosexual, white (i.e., do not study immigrants as some of them might be undocumented or do not study black workers because some of them may fear reprisal from their white employers), and male (i.e., women might report domestic violence).[35]

A British researcher argued that the complexity of consent forms restricted research to "the articulate literate well . . . and therefore excludes many people whose needs should be captured."[36]

Any mention of sex set off alarms. Patricia Adler and Peter Adler watched in dismay as IRBs put so many conditions on studies of gay teenagers, public sex, and sexually transmitted diseases that graduate students abandoned the topics; one student left the university in disgust.[37] Tara Star Johnson, who wanted to talk to teachers about the sexual dynamics of their classrooms, was told that her project was so risky that she would not be allowed to record her interviews. After she protested, the board decided that her project posed only minimal risk.[38] A historian interviewing middle-aged gays and lesbians was asked to invent pseudonyms and destroy her tapes and transcripts.[39]

Such hypersensitivity to controversy threatened some of the most important research conducted by social scientists. Anthropologist Scott Atran, for example, sought to interview failed suicide bombers in order to learn how to negotiate with other potential terrorists. Since the bombers were in prison, however, the University of Michigan's IRB told him that they could not give consent, and he would likely never receive permission to interview them. The IRB did grant permission for Atran to speak to jihadis still at large, but only under arbitrary conditions that restricted the questions he could ask and the results he could publish, and even this permission was capriciously withdrawn. "How is anybody in academia ever going have as much as possible to offer in this whole mess," Atran asked, "if no one can even talk to the people most involved?"[40] In Britain, Robert Dingwall wanted to survey staff in 350 hospitals to find out why some hospitals illegally reused single-use medical devices. An ethics regulation would have required him to get "approval from each site, potentially generating about 1600 signatures and 9000 pages of documentation" and "would also have required my colleague to undergo around 300 occupational health examinations and criminal record checks." Unable to meet these demands, Dingwall and his colleague scaled back their potentially life-saving work.[41]

In some cases, university administrators claimed powers far beyond those envisioned by the federal authors of the regulations. The OPRR had nearly suspended research at UCLA in the 1990s, and a decade later the university's Office for Protection of Research Subjects was still wary of letting any research go without review.[42] That office announced in 2007 that it had "the sole authority to determine whether an activity conducted by UCLA faculty, staff, or students (or conducted on UCLA students) meets the regulatory definition of 'human subjects research' and therefore requires IRB review and approval or certification of exemption from IRB review." As a result, "investigators who intend to conduct activities that might represent 'human subjects research' must submit a description of the proposed activities." Not only were researchers asked to submit protocols that the researchers believed *did* fit the regulatory definition, but they were now asked to submit protocols that, they guessed, *might represent* regulated research to some unidentified person.[43] The implications of such a sweeping claim of jurisdiction became clear in 2008, when UCLA gave its researchers specific instructions on how to apply for permission to quote (or perhaps merely to read) a published letter to the editor or a blog.[44]

In other cases, IRBs ignored the regulations' clear statement that only risks to the individual subjects of research—not the implications of the findings—should be considered. Geographer Matt Bradley had the misfortune to submit a proposal for a documentary film in the 1998–1999 academic year, just as his campus IRB was beginning to impose oversight on qualitative research. Because such oversight was so new, neither the IRB nor Bradley's colleagues knew much about the process, and he almost escaped review on the grounds that he was conducting journalism, not research. But the IRB staff told him to submit a proposal, and, after months of delay, his project was prohibited on the grounds that "there is risk that people in the community might be upset about the portrait that has been painted."[45]

The delay faced by Bradley became commonplace. At Northwestern University, "the time required to pass review increased from usually around forty-eight hours for social science reviews to what could be months for even the most routine projects."[46] Such delay often killed projects outright, as faculty lost the chance to do research over the summer, and students could no longer expect to complete projects within a semester.[47] At another university, graduate students regularly blamed the IRB for delaying their graduations by whole semesters.[48] Even for faculty, delays could kill projects if grant money or participants disappeared.[49]

Perhaps most frustrating to researchers is the wild inconsistency of responses to similar protocols, leading one scholar to call IRB review "a game . . . using a ouija board."[50] Education professor Jim Vander Putten, himself an IRB chair, sought to interview faculty and staff at five research universities. Since the project had already been approved by one IRB, Vander Putten was shocked when IRBs at the other schools to be studied demanded petty and inconsistent changes, with one board insisting that consent forms be written in the past tense, and another forbidding that practice. Complying with all the demands delayed the project by months. "For untenured faculty," he warned, "these delays can present formidable obstacles to meeting institutional expectations for scholarly productivity leading to tenure and promotion."[51]

Music educator Linda Thornton and a colleague at another university wanted to survey music education majors at the twenty-six top university programs to ask why they had chosen music education as a profession—an innocuous question that should have been granted swift exemption from review. Instead, an IRB forbade the researchers from interviewing students at their own institutions, and required them to seek permission from the IRBs at the remaining twenty-four universities they wanted to study. Nine of the twenty-four accepted the proposal as approved by Thornton's IRB, including one that noted it had a reciprocity agreement in place. Of the remaining fifteen, several imposed burdensome requirements, ranging from small changes in the informed consent letter (which then needed to reapproved by the original IRB) to the requirement that the instructor at the local institution, who was just going to distribute and collect questionnaires, be certified in human subjects research. Application forms varied from two pages to eight; at least one IRB demanded to know the exact number of music education majors in every school to be surveyed. The result was that the researchers dropped many of the schools they hoped to study, cutting their sample from several thousand to 250.[52]

Even within a university, similar projects may get varied responses. "We routinely are granted permission to give extra credit to students for their study participation," wrote some researchers.

> But once we were told that offering extra credit is unduly coercive to students, and we should instead, like our Psychology Department, make participation in research a requirement of our courses and thus part of students' grade in the course. Incidentally, our Psychology Department does not in fact have such a policy . . . In another example, we once submitted an informed consent form that had previously been approved. It was virtually identical to the previous one except for minor word-

> ing and title changes. The IRB came back with eleven major changes required before it could be approved.[53]

Another noted that if two students submitted similar projects, one might be required to "go through a tedious consenting process taking as much as four to five minutes of the beginning of telephone interviews (with the busy elites at the other end chafing at these long and unnecessary prefaces to the first question), while another student researcher is not required to secure any kind of consent at all."[54]

In the saddest cases, researchers censored themselves before even submitting proposals. One Northwestern University scholar wrote that in the 2000s, "my research became more theoretical in large part because of IRB requirements . . . I no longer interview people during my trips abroad and try to limit the data gathering to passive observation or newspaper clippings."[55] Another researcher despaired of being allowed to interact with teenagers in online chatrooms; he had to be content with observational work.[56] "The whole process has now become so cumbersome and depressing that I am seriously considering how much longer I intend to remain doing this kind of work," complained a British social scientist who studied health care. "I'm not kidding—health-related research is a soul-destroying, unhealthy business these days."[57]

Students were especially prone to abandon projects before they had begun. In Britain, professors steered graduate students away from health-related empirical research because the approval process took too long.[58] In the United States, a graduate student in education wanted to speak with HIV-positive minors but—intimidated—asked only to speak with adults who interacted with those youth in schools.[59] Another bright student who had been involved for years in AIDS activism shied away from proposing an ethnographic study of her fellow activists. Having seen the IRB difficulties faced by students ahead of her, she proposed "a statistically significant, but dull, survey of the relationship between healthy eating habits and extracurricular activities of college students." At the University of North Carolina at Chapel Hill, a professor who once asked students to explore such topics as binge drinking, date rape, and academic dishonesty—all of them designed to excite student interest—gave up after too many encounters with the IRB and limited himself to "bland topics and archived records."[60] An anthropologist, who found her own IRB unable to offer significant ethical advice, worried about graduate students beset by "IRB fatigue."[61]

Some scholars feared that such self-censorship, in the aggregate, could weaken whole fields of social inquiry. In Canada, graduate students in anthropology continued to interact with people through interviews, but they increasingly abandoned participant observation, formerly one of the basic tools of their discipline. On discovering this trend, two scholars warned of a "pauperization of anthropological research," especially as a new generation grew up with "the medical model of research."[62] In their 2004 biography of Laud Humphreys, the sociologists John Galliher, Wayne Brekhus, and David Keys blamed IRBs for the emergence of a "tame sociology" in which "sociologists find it much more difficult to study human behavior, and often merely investigate what people say about their behavior."[63] Likewise, Mary Brydon-Miller and Davydd Greenwood feared that "while providing protection," committee review could also "have the effect of making social research largely impotent in terms of addressing issues of real importance."[64]

Most of these horror stories are written from the point of view of the frustrated researcher. A few scholars—most notably anthropologist Maureen Fitzgerald and sociologist Laura Stark—decided to see what the process looked like from the other side of the door, and they gained permission to watch IRBs in action. What they saw bore little resemblance to the procedure laid out in the *Belmont Report*, which called for "systematic, nonarbitrary analysis" and "the accumulation and assessment of information about all aspects of the research" that would render "the assessment of research more rigorous and precise, while making communication between review board members and investigators less subject to misinterpretation, misinformation and conflicting judgments." Fitzgerald and Stark both witnessed assessments that were nonsystematic, arbitrary, and vague, leading to highly conflicting judgments. Fitzgerald and her colleagues found that IRB members felt free to base decisions on worst-case scenarios or personal anecdotes with the reliability of "urban myths or contemporary legends." At all three IRBs she observed, Stark saw members judging proposals based on the proportion of spelling and typographical errors in the proposal.[65]

And what of the other side of the balance sheet? Where are the accounts of successful IRB interventions in social science and humanities research? Most defenses of IRBs cite only noninterference. One anthropologist who boasted that the "IRB is working very well" at her university also noted that "I have not had to alter my research for IRB ever."[66] Perhaps there is an argument to be made that a do-nothing IRB is preferable to no IRB at all, but few scholars seem willing to make it. IRB review does put more sets of eyes on a project, and boards may sometimes catch errors, such as a set of instructions that does not match the questionnaire being distributed.[67] One IRB member insisted that a proposal "emphasize its intention to alter names of elementary school student participants for publication. I know that some of these changes might seem bureaucratic, trivial, even annoying to faculty members," the scholar reported, "but they underscore the importance of doing research by ethical means."[68] Yet it is rare to find a scholar in the humanities or social sciences who credits an IRB with helping to clarify significant ethical questions.

That's not to say it doesn't happen. In one survey, 32 percent of British social researchers who had gone through a National Health Service ethics review reported that the experience had improved their projects. In the same study, however, 51 percent of respondents reported that ethics approval imposed changes for the worse, suggesting that committee review overall does more harm than good.[69] Oral historian John Willard credited the IRB at the University of Maryland, Baltimore County, for placing sensible conditions on three projects: "interviews with a Lebanese Druze Muslim man who still could be prosecuted for what he was going to discuss; the project that interviewed elderly homosexual men about homosexual culture during 1940–50s; and a project involving end of life issues with interviewees being interviewed in hospice care."[70] Likewise, sociologist Peter Moskos was grateful to Harvard's IRB for requiring him to announce his ethnographic intentions to his classmates in a police academy. (Nevertheless, he remained an IRB skeptic.)[71] British sociologist Adam Hedgecoe was glad to see an ethics committee forbid a nurse from asking members of her team about their attitudes toward performance pay. He and the committee thought the study potentially coercive; the researcher felt her work was being stifled and filed a complaint.[72] But even counting such an event as both a horror story (for the researcher) and a success (for the committee), it is vastly easier to find examples of abuses by IRBs than abuses by social scientists. And rarer still are abuses that could have been prevented by prospective review.

By 2007, even the leaders of the IRB establishment showed some awareness of social researchers' discontent. Deborah Winslow of the National Science Foundation—a Common Rule signatory agency—lamented that "the actual functioning of some IRBs is so at odds with common sense and researcher realities that serious, socially responsible researchers come to feel they have to ignore their IRB and break the law."[73] In 2008, the newsletter of the Association for the Accreditation of Human Research Protection Programs conceded that "many behavioral and social scientists feel constrained by a system that seems tilted toward biomedical research and, therefore, neither understands nor reflects their concerns."[74] And in response to questions from a *New York Times* reporter, the head of the OHRP, Bernard Schwetz, conceded that his office had failed to offer clear guidelines for the review of nonmedical research. Schwetz promised "a lot of examples" and said the OHRP "will give more guidance on how to make the decision on what is research and what is not."[75] In April 2007, Schwetz reiterated his promise in a letter to historian Don Ritchie, pledging that the "OHRP will seek public comment on a draft version of the guidance document before it is finalized."[76] But Schwetz retired that September, and no document appeared by his promised year-end deadline—or by the end of the following year.

The OHRP's broken promise continued a tradition, dating back to 1966, of empty assurances that attention to the social sciences was just around the corner. But after four decades of IRB policy, it was very hard for figures in authority to abandon their presuppositions, as well as for social scientists to expect policies that paid attention to their interests.

ACCOMMODATION

Many social researchers, including leaders of social science organizations, called for scholars to cooperate with IRBs. These scholars did not deny serious problems with the way the regime had operated since the late 1990s, and they suggested significant changes in that regime. However, they insisted on maintaining the *Belmont Report*, the Common Rule, and the existing basic structure of IRBs. They sought reform, not revolution.

Advocates of accommodation tended to share five basic premises. The first premise is that social research is connected—methodologically, ethically, and legally—to the medical and psychological research on which IRB review was based. One communication scholar, an IRB member, proclaimed, "I need not detail here the clear excesses of past research studies (such as the Milgram research); ideologically, the need for an IRB is firmly established." In this case, a single reference to a forty-year-old psychological experiment is used to justify IRB oversight of dramatically different research methods.[77] Anthropologist Stuart Plattner wrote that "no one should ever be hurt just because they were involved in a research project, if at all possible," a noble statement for medical researchers, but not one that everyone would apply to the social sciences.[78]

Conversely, a second premise is that social research is distinct from journalism, which operates without IRB review outside, and often inside, the university. In 2004, for example, Charles Bosk and Raymond De Vries offered two distinctions between ethnography and journalism: first, that "journalism is not supported by public funds," and second, that journalists "lack what are often thought of as the prerequisites of a professional occupation—long adult socialization in a specialized body of theoretic knowledge."[79] Law professor James Weinstein offered a third distinction: that social science simply wasn't important enough to merit the constitutional protection once claimed by Ithiel de Sola Pool. The First Amendment would cover interviews concerning "attitudes towards homosexuality, abortion, or the war in Iraq," he argued, but, "unlike the typical journalistic interview or survey, many social science interviews and surveys will not contribute to democratic self-governance."[80] Plattner concurred, suggesting in 2006 that while "the journalist has a mandate from society to document contemporary reality . . . social scientists have no such mandate; we document reality to explain it. Our audience is professional, and society gives us no protection in the First Amendment."[81]

None of these arguments is very carefully advanced. Bosk and De Vries overlook the question of university schools of journalism, which do receive public funds and do impart some body of knowledge, if not a theoretic one. As for Weinstein's suggestion that the typical social science project contributes less to democracy than the typical journalistic interview or survey, one can only assume that he had never compared an issue of the *American Sociological Review* to one of *People* magazine—or read the works of Woodrow Wilson or Daniel Patrick Moynihan. And Plattner assumed that social scientists never share their research findings with a general public, which is hardly the case.

A third premise is that federal agencies, university research offices, and IRBs are staffed and managed by well-intentioned people who wish to promote ethical research. "IRB members are not those folks who are looking to thwart your study," promised epidemiologist J. Michael Oakes in 2008. "They are peer researchers who have a job to do."[82] Plattner agreed: "With good faith efforts from all participants in the human subjects research system, we can pursue the goals of advancing research while minimizing exposure to risk of harm."[83]

A fourth premise is that these well-intentioned people should be able to succeed because federal regulations and guidance are relatively flexible. After all, nothing in the rules or guidelines explicitly prohibits any type of research. As Plattner argued, "the regulations seem complex and daunting on first reading, but in fact they allow a fair amount of flexibility. That assumes they are administered by people with common sense who understand that research is a public good which should not be impeded without a clearly defined, reasonable risk of harm."[84] Likewise, Felice Levine and Paula Skedsvold claimed that "creative use of the flexibility within the current system might resolve some of the pressing concerns of social and behavioral science investigators while ensuring adequate oversight of research involving human participants."[85] In 2005 Kristine Fitch assured her fellow ethnographers that the OHRP "can be notified of hypervigilant regulation on the part of a local IRB. They can sanction IRBs for over-interpretation or misapplication of regulations when there is evidence that such is the case."[86] However, she did not claim that the OHRP had ever imposed such a sanction. Indeed, these accounts generally ignore the long history of federal enforcement efforts, dating from the OPRR's crushing of the University of California, Berkeley's proposed system in 1974 to the OHRP's shabby treatment of oral historians in 2003.

The final premise is that the regulations, however flawed, are unlikely to change, so it is best to learn to live with them. As Bosk explained in 2004, "prospective review strikes me as generally one more inane bureaucratic requirement in one more bureaucratic set of procedures, ill-suited to accomplish the goals that it is intended to serve." Nevertheless, he continued,

> prospective review, flawed a process as it is, does not strike me as one social scientists should resist. After all, we agree with its general goals: that our informants should not be subject to untold risks, that they be treated with decency, that their confidentiality and anonymity be safeguarded, when feasible. Given this, we should not waste our energies resisting a process that has an enormous amount of both bureaucratic momentum and social consensus behind it. Instead, we should focus our energies on reforming and revising procedures; we should fix the system where it is broken.[87]

Anthropologist Patricia Marshall shared such fatalism. In an essay published in 2003, she acknowledged that "misapplications of the Common Rule and inappropriate requests for revisions from IRBs can have a paralyzing effect on anthropological research. Moreover, it reinforces a cynical view of institutional requirements for protection of human subjects, and it uses scarce resources that would be better spent on studies involving greater risks for participants." Yet she insisted that "regulatory oversight by IRBs is a fact of life for scientific researchers. Anthropologists are not and should not be exempt."[88] Plattner agreed: "It is a waste of time to think of changing the regulations, which, after all, took ten years to craft."[89] At the local level, some scholars accept the approval process simply because it's easier to comply than resist. "There is usually some trivial correction or change that the IRB requests," reported an anonymous scholar. "I think it makes them think that they are doing a good job if they can find something wrong that needs to be changed . . . I make the change and the study is approved. The changes do not seem to really protect human subjects, but they are easy to make and it is not worth arguing about. It is easier to just go along and play the game."[90] Such arguments did not necessarily claim that IRBs *do* safeguard the rights and welfare of participants in social research, only that their general goals are noble. Levine and Skedsvold argued that mere miscommunication created "frustration and skepticism in a system that could *potentially* work quite well if transformations are made" (emphasis added).[91]

Some federal officials outside of the OHRP tried to achieve such transformations by restricting the reach of IRBs without amending the federal regulations. In 1999 the working group that had recommended the exclusion of history from IRB review also called on IRBs "to adopt creative administrative and other means to reduce administrative burden and maximize attention to the most important ethical issues."[92] The National Science Foundation (NSF)—the Common Rule signatory most likely to sponsor social science research—also offered interpretations of the regulations that were considerably more flexible than the OHRP's own guidance. Around 2002, the NSF published a set of answers to frequently asked questions about the federal regulations. It noted, for example, that "when the subject can readily refuse to participate by hanging up the phone or tossing out a mailed survey, the informed consent can be extremely brief," and that "in certain circumstances, persons are not in a position to decide whether to consent until after their participation."[93] The 2008 version of the NSF document included arguments that social scientists had been making for years, noting, for example, that "in most ethnographic projects a request for a written, formal consent would seem suspicious, inappropriate, rude and perhaps even threatening." The NSF also directly challenged common IRB demands:

> Detailed recitation of irrelevant information demeans the communication and is slightly insulting. People are capable of deciding whether to participate in surveys and ethnographic research. Assurances that there are no risks and descriptions of measures taken to assure confidentiality can be irrelevant, irritating, misleading, and may not decrease the risk of harm.[94]

The problem for researchers is that such guidance only applies to research directly sponsored by the NSF. In the absence of such sponsorship, IRBs still turn to the OHRP for direction.

Two quasi-official advisory commissions recognized the problem of overregulation while hoping to work within the present system. The first grew out of the National Human Research Protections Advisory Committee (NHRPAC), established in 2000 by Donna Shalala, secretary of Health and Human Services. In February 2001, the committee established a Social and Behavioral Sciences Working Group—cochaired by Felice Levine—that included psychologists, sociologists, educational researchers, and a medical anthropologist, but no historians, political scientists, journalists, or folklorists.[95] When NHRPAC's charter expired in 2002, the members of the Working Group decided to keep meeting.[96] In 2003, with NIH funding, this group held its own conference at the Belmont Conference Center.[97]

Between 2002 and 2004, the Working Group issued a series of papers on such specific topics as evaluating risk and harm and handling research declared exempt under the regulations. The papers acknowledged researchers' frustrations with IRBs, noting, for example, that IRBs "have too frequently operated unaware of the nature of social and behavioral science research involving human subjects, the likely risks and harms associated with such research, and the best procedures for protecting subject populations involved in such research."[98] They also recommended moderate structural changes to the review process, such as the inclusion of more social and behavioral researchers on IRBs and the delegation of some review to academic departments.[99]

Yet the Working Group shied away from any consideration of ethical principles as sweeping as the original *Belmont Report,* nor did it address such big questions as whether prospective review of research was the best way to ensure ethical social science, or how researchers could protect themselves from capricious yet all-powerful IRBs. As Raymond De Vries, who attended the 2003 Belmont conference reported, "one more working group's recommendations are more likely to fuel rather than to extinguish the flames of discontent."[100]

Another quasi-official effort at reform was the work of the awkwardly named Panel on Institutional Review Boards, Surveys, and Social Science Research, convened in 2001 by the National Research Council. Like the Working Group, the Panel included scholars in sociology, psychology, medical anthropology, and survey research, but not in history, political science, folklore, or journalism. This makeup was particularly ironic, given the Panel's eventual recommendation that "any committee or commission established to provide advice to the federal government on human research participant protection policy should represent the full spectrum of disciplines that conduct research involving human participants."[101] And like previous commissions, the Panel failed to define social science or to explore the variety of disciplines within that category. Thus, while the panel's name referred only to surveys and social science research, its final report, issued in 2003, was entitled *Protecting Participants and Facilitating Social and Behavioral Sciences Research.* So was it a panel on social science, or on behavioral *and* social science? And what types of research fit under each label?

Protecting Participants noted real problems with the present system. Like many earlier reports, it suggested that IRBs were spending far too much time reviewing research that posed little risk, inconveniencing researchers while burdening IRB members.[102] It conceded that some IRBs might "request changes in research design that compromise the scientific validity of the study without necessarily increasing protection for participants."[103] Nor did it offer empirical evidence that IRBs were protecting anyone. Yet the Panel did not investigate alternatives to IRBs, or the possibility—raised by the National Bioethics Advisory Commission—that some forms of research "might be better regulated through professional ethics, social custom, or other state and federal law."[104] The Panel took the *Belmont Report* principles and applications as a given, neither questioning their relevance for social science nor investigating alternative ethical standards.[105] And rather than call for any changes in the regulations, it sought to "encourage the use of the flexibility that is currently possible according the Common Rule."[106] Like the Working Group, the Panel called for the OHRP to issue detailed guidelines on a number of topics without examining the quality of the guidance issued in the past.

Protecting Participants ended with a plea for more research. After tracing all the studies of IRB operations from 1973 to 1998, it concluded that "there is astonishingly little hard information about the operation of the IRB system . . . and how IRBs are interpreting the Common Rule," and it called for the federal government to collect regular data and to sponsor in-depth studies of IRBs.[107] Though the OHRP did not respond to this suggestion, in 2006 psychologist Joan Sieber—a longtime veteran of IRBs and ethical debates—founded the *Journal of Empirical Research on Human Research Ethics* to collect studies of the risks and benefits of various types of research, thus building an "evidence-based ethics" in place of the guesswork so common to IRB review.[108] Such studies punctured some of the assumptions common to IRBs, such as the imagined dangers of interview and survey research.[109] But though occasionally critical of individual IRBs, the journal's editorials counseled social scientists to work within existing regulations.[110]

Taken together, the Working Group reports, *Protecting Participants*, and other accommodationist documents imagined an IRB regime in which scholars of all stripes would volunteer to serve on IRBs or departmental ethics committees. Perhaps they would be "compensated with release time from teaching and recognition in promotion decisions."[111] And once on the board, they would benefit from constant, wise guidance from the OHRP, which would abandon its compliance-oriented enforcement of the past. The OHRP, in turn, would be advised by a federal commission unlike any yet formed, since it would be composed of representatives of all scholarly disciplines subject to regulation. At every level, officials and board members would consult plentiful empirical research on the conduct of social science and the workings of IRBs. The crackdown of 1998 would be forgotten, the horror stories would vanish, and though IRBs would still wield enormous power over individual researchers, they would use that power only for the common good. And all of this would take place without changing a word of the Common Rule or the *Belmont Report*.

By the late 2000s, proponents of this vision conceded that it remained elusive. As Felice Levine wrote in 2007, "despite considerable discussion over the last five or more years . . . the absence of any visible net improvements in advancing the sound review of social and behavioral sciences research has impeded some important research, limited training opportunities, and unfortunately produced disaffection on the part of too many researchers, while also distracting IRBs from matters needing their time and attention."[112] Faced with these same facts, other scholars concluded that IRB review of social research was inherently flawed.

RESISTANCE

While some scholars in the social sciences and humanities embraced IRB review, others resisted it as strongly as had Ithiel de Sola Pool in the 1970s. While resisters often looked at the same facts as accommodationists, they tended to reject one or more of the premises underlying the accommodationist approach.

First, they often embraced methodological and ethical frameworks closer to those of investigative journalism than to medical research. Ethnographers are particularly insistent that review processes designed for experiments cannot fairly judge open-ended investigations. Sociologist Jack Katz complained that "ethnographers [often] find it impossible to seek preauthorization for observations and interviews, no matter how sincere the will to comply."[113] Martin Tolich and Maureen Fitzgerald lamented that "in our own research projects, as well as Fitzgerald's extensive study of ethics committees in five countries . . . we have yet to find an ethics committee that reflects qualitative epistemological assumptions."[114]

As for ethics, IRB skeptics embraced the idea that good research could hurt. As historian Linda Shopes argued in 2007, "historians' deepest responsibility is to follow the evidence where it leads, to discern and make sense of the past in all its complexity; not necessarily to protect individuals. In this we are more like journalists and unlike medical professionals, who are indeed enjoined to do no harm."[115] Likewise, anthropologist Richard Shweder argued that scholars in the social sciences, humanities, and law should follow the path of Socrates, notwithstanding that he was put to death "for being annoying and for asking questions that powerful members of the community did not want asked."[116] English professor Cary Nelson (later an AAUP president) argued that IRB members themselves saw no principled distinction between scholarship and journalism; they simply dared not take on the press.[117]

Resisters also rejected the second and third premises of IRB accommodationists: that IRB members and staff are well-intentioned people devoted to the protection of research participants, and that regulations and OHRP guidance do not force misbehavior. Rather, resisters see a system of incentives that make IRBs less concerned with protecting the subjects of research and the freedom of researchers than with protecting universities from lawsuits and the loss of federal funds, or blocking research that might challenge established political views.[118] As Murray Wax testified in 2000, "there is a natural tendency for institutions . . . to protect elite access to federal funding . . . Disregarding the actual ethical issues, the regulators wish to safeguard the $50 million project by subjecting the $50,000 projects to project requirements that are irrelevant."[119] IRB members occasionally admitted such priorities. At the University of Memphis, an IRB member told sociologist Carol Rambo, "I don't give a tinker's damn about your human subject. This is all about liability. The University doesn't like lawsuits or anything that could tarnish its image."[120] An IRB chair at the University of Chicago boasted in print that IRBs "can forestall the public image problems and protect the institution's reputation by weeding out politically sensitive studies before they are approved."[121] While some of these incentives have been in place since 1966, the 1990s crackdown intensified them. A 2007 article found that "the IRB's over-riding goal is clear: to avoid the enormous risk to the institution of being found in noncompliance by OHRP. IRBs . . . must see everything around them as a potential source of catastrophic risk."[122] IRB members, Nelson wrote in 2004, "are afraid of the federal regulators, just as faculty members are afraid of their IRB."[123]

Even less-charitable explanations exist. Some suspected individual self-interest to be at play. As IRB administration became a career in itself, administrators could only gain resources—salaries, promotion, conference travel—if IRB jurisdiction expanded and if they could justify their budgets with signs of activity. In the words of Charles Bosk, "the functionaries who staff the new bureaucracies of virtue are able to create the impression of efficiency by shifting burdens directly to researchers and their staff. Action on proposals, a measure of administrative activity, occurs when proposals are returned to researchers for reasons no more serious than incorrect font size, incorrect pagination, or other niggling matters."[124] Law professor Scott Burris bemoaned a system of "rituals of compliance and self-congratulatory gestures conducted to no great effect under the stern gaze of self-appointed virtue experts."[125]

As for the faculty members on the IRBs, resisters suspected them of imposing their own methodological biases rather than protecting research participants. "The IRB at my university consists of primarily post-positivist, quantitative researchers," complained one qualitative researcher. "As I read through my first deferral letter I realized the IRB simply did not understand my research methods."[126] Others suggested that IRBs act irrationally, victims of "moral panics."[127] Even some IRB members conceded the point. As UCLA anthropologist and IRB member Frederick Erickson put it, "while I was behind the curtain, it seemed to me that our board was quite reasonable . . . Now, I've got a project of my own in expedited institutional review," he added, "and it's being jerked around in ways that make my blood boil."[128] But angry researchers—especially junior faculty and graduate students—posed little threat to IRB members and administrators. IRBs would always face greater danger from being too lax rather than being too strict.

For critics, IRB review was a barrier, rather than an aid, to true ethical reflection. As historian and ethicist Alice Dreger complained in 2008, researchers "often think that, once they've gotten IRB approval and 'consented' their subjects, their ethical obligations are done . . . Even if they began the IRB application experience, in graduate school, with real ethical reflection, they've lost it three or four rounds in."[129] Charles Bosk (having apparently grown more frustrated with IRBs since his 2004 essay) wrote in 2006, "The most serious defect of the current regulatory system is that the requirements of policy reduce and trivialize the domain of research ethics."[130]

Finally, naively or not, resisters saw real possibilities for changing public policy, particularly the Common Rule regulations. Even official bodies took this view. In 2000, HHS's Office of Inspector General argued that while trying to get seventeen federal agencies to agree to change the regulations was difficult, a legislative change might make sense.[131] In its final report, the National Bioethics Advisory Commission concluded that too much research was subject to oversight, and it called for a new federal office that would "initiate a process in which representatives from various disciplines and professions (e.g., social science, humanities, business, public health, and health services) contribute to the development of the definition and the list of research activities subject to the oversight system."[132]

Scholars wishing to resist the IRB system had various options. The simplest form of resistance was to do one's research without informing the board.[133] This was easiest for senior scholars. In 1998 sociologist Howard Becker, by then nearing the end of his career, ignored colleagues who warned him to seek IRB approval before sending students to talk to passengers at a bus station. He figured that the IRB enforcers "are certainly not about to fuss with senior people."[134] In 2005, sociologist Edna Bonacich described similar defiance. "I personally think that the main mission of the IRB is to protect the university from being sued, but it is veiled with the language of 'looking after' the well-being of research subjects," she explained. "I simply do not deal with the IRB . . . I know that I am violating university regulations, but I believe there is a clear, pro-business aspect to this policy." Yet self-exclusion is not for everyone. As Bonacich acknowledged, an established, tenured scholar takes less of a gamble than a graduate student, postdoctoral fellow, or untenured professor. For such junior scholars, she suggested confronting the IRB or "disguis[ing] your political intentions under the language of scientific research."[135]

Indeed, researchers learned to disguise not just political intentions, but all sorts of plans and modifications as they pursued their work. Ethnographers, in particular, learned to promise one thing to IRBs, then forget those promises once they gained approval. In 2000, sociologist Ann Swidler complained that IRBs "turn everyone into a low-level cheater . . . I'm not sure there is much relationship between what people agree to do and what they do in the field."[136] Kevin Haggerty agreed, writing in 2004 that "we have reached the point where breaking many of the rules imposed by [research ethics boards] would not in fact result in unethical conduct—if ethics is conceived of as anything beyond simple rule-following."[137] An Australian ethnographer knew that some questions she might ask could distress her narrators, but she also knew that confessing as much to the ethics committee would produce inappropriate demands that she offer professional counseling to anyone who "became a little teary during an interview." She kept the possibility of distress off her ethics application, reasoning, "we can't be too broad/honest in our thinking when filling out these forms."[138] IRBs themselves may take part in this dishonesty, "tacitly agree[ing] to orchestrate their exchange of information in such a way that potential elements of an inevitably gray factual world can be framed using language that may pass IRB muster," even if that language disguises the researcher's true intent.[139]

Other scholars went further, deliberately deceiving the boards. In the 1999–2000 academic year, the Rutgers IRB got in a spat with the university's historians, who were planning to interview former members of the New Jersey state legislature. After months of wrangling, the project got IRB approval, but only after the researchers promised to avoid stressful questions—a promise they did not intend to keep.[140] Another scholar routinely began research while the application was still pending, explaining, "I don't date any of [my] data collection so no one can prove I collected it before the approval arrived."[141] Yet another left collaborators' names off the IRB application, lest they be required to complete useless ethics training.[142] In surveys, between 8 and 20 percent of nonmedical researchers admitted having avoided compliance with some human subjects requirements.[143]

Other scholars resisted more publicly. Some sought to free themselves of IRB review by persuading their universities to take advantage of the fact that federal law and regulations only apply to research directly funded by the federal government, unless a university promises to review unfunded research as well. By January 2006, 174 American universities and colleges, some of them among the most prestigious in the nation, had filed federal assurances which did *not* pledge review of unfunded research. In practice, though, those universities' administrations continued to insist on IRB jurisdiction over unfunded work.[144] This helped universities reduce their exposure to the OHRP sanctions, but it did nothing for researchers seeking more freedom. Moreover, several states had passed laws requiring IRB review of all human subjects research, regardless of its funding source or the affiliation of the researcher. Were scholars in one of those states to escape IRB review as a federal requirement, they would still—in theory—face it simply by the fact of their residence.

Law professors, as well as legal-minded colleagues from other disciplines, began pondering legal challenges to the whole IRB regime. The legality of IRB rules had been debated in late 1970s, with Ithiel de Sola Pool and others declaring that the proposed regulations of 1979 were unconstitutional. But John Robertson—a law professor who had acted as a consultant to the National Commission—argued that funding agencies and universities enjoyed great leeway on the conditions they attached to grants or contracts, and no legal scholars challenged his conclusions.[145] Other scholars revisited the question in the 2000s and found the regulations to be "constitutionally vulnerable," though they conceded that a legal challenge was far from certain to succeed.[146]

By the mid-2000s, IRB skeptics from a variety of disciplines had begun to find each other, just as they had in 1978 and 1979. In April 2003 scholars from law, journalism, anthropology, English, psychology, and other fields at the University of Illinois at Urbana-Champaign hosted a conference on IRB oversight of nonbiomedical research. The result was the "Illinois White Paper," a report that blamed IRBs' fear of lawsuits or further federal shutdowns for the extension of oversight over ethnography, journalism, and oral history, fields that "pose virtually no risk to the subjects," or at least not to subjects who don't deserve to be held accountable for misdeeds.[147] This "mission creep," the *White Paper* warned, not only interfered with the freedom of social scientists to do their work, but it also hampered IRBs in their main task of protecting participants in clinical, biomedical studies. The term mission creep became a popular new shorthand for the extension of IRB powers beyond biomedical research, but most of the arguments in the *White Paper* could have been made in 1978, or even 1974.

The durability of old critiques of IRBs was even more apparent in a 2006 report by the American Association of University Professors, prepared by a committee chaired by Judith Jarvis Thomson—who had also been the chair of the 1981 AAUP subcommittee that had complained that the exemptions in the new federal regulations were inadequate. This new report was written very much in the spirit of Ithiel de Sola Pool's revolt twenty-five years earlier, noting that the regulations "give no weight at all to the academic freedom of researchers or to what the nation may lose when research is delayed, tampered with, or blocked by heavy-handed IRBs." It described IRB worries about distressing questions to be "an unpardonable piece of paternalism where the subjects are adults who are free to end their participation at any time, or to refuse to participate at all."

The AAUP report even offered a version of the Pattullo formula, suggesting that the regulations be amended to state "that research on autonomous adults whose methodology consists entirely in collecting data by surveys, conducting interviews, or observing behavior in public places, be exempt from the requirement of IRB review—straightforwardly exempt, with no provisos, and no requirement of IRB approval of the exemption."[148] In short, social scientists now found themselves back where they were in 1978, only with a greater awareness of how readily regulators might interpret exemption clauses in a way that nullified them. The report recognized that talk is cheap: "Complaints published here and there over the years have accomplished little beyond generating an angry and deeply dismaying literature." It suggested a joint effort of scholarly and educational associations—led by the AAUP itself—to change the regulations.

This resistance movement nevertheless remained a pale shadow of the 1977–1980 effort. While law professors were willing to craft theoretical challenges to the constitutionality of the regulations, none seemed eager to file a lawsuit. No one in the resistance camp was as chummy with White House officials and members of Congress as Pool had been. And whereas the earlier movement had united a dozen scholarly organizations in a mostly solid front in favor of Pattullo's exclusion clause, thirty years later key groups—including the American Anthropological Association, the American Sociological Association, and the Consortium of Social Science Associations—were silent or willing to live with the IRBs. With social scientists divided about the basic legitimacy of IRBs, university administrators, regulators, and legislators had little reason to choose one side over the other.

ALTERNATIVE MODELS

Despite the often bitter disputes between advocates of accommodation and those of resistance, scholars on both sides have suggested alternative models that would maintain some oversight of research, but with very different procedures and, presumably, different outcomes.

One proposed reform would be to demand that IRBs adopt what Jack Katz has called a "culture of legality," with three characteristics. First, IRBs and university administrators would "never command the impossible." Second, they would not set policy without consulting those affected. And third, they would publish their decisions, along with the reasoning behind those decisions, so that researchers at any university could use them as precedents to be cited in their own applications.[149] IRBs around the world could, potentially, build a common body of knowledge, thus eliminating the ouija-board arbitrariness that so antagonizes researchers and deprives research participants of meaningful protection. Though Katz wrote as a fierce critic of IRBs, his ideas were shared by at least some IRB advocates. For example, in 2000 Harold Shapiro—the chair of the National Bioethics Advisory Commission—floated the idea of developing a "sort of common law" through "publicly available" decisions.[150] In 2008, in a generally accommodationist essay, Felice Levine and Paula Skedsvold called for IRBs to exhibit "procedural justice," consisting of "respectfully considering group members' views, treating group members with dignity, and demonstrating neutrality in decision-making processes."[151] Like Katz, they hoped that such responsiveness and transparency would reduce the hostility between researchers and review boards.

A second proposal for change imagined the decentralization of review, so that most review would take place at the level of the individual department or research unit. As Levine and Skedsvold wrote, "decentralization allows for more relevant expertise, more interaction around ethical issues and problem solving, and a more educative approach to review and approval of protocols."[152] Something of the sort was in place at Macquarie University in Australia, where, as anthropologist Greg Downey described it, "members of the committee are clustered so that color-coded sub-groups do the preliminary and most serious review of applications for which they have special expertise."[153]

Finally, a third proposal would replace board review of highly specific protocols with a system under which scholars could be certified to conduct certain forms of research. Most notably, in 2006 the University of Pennsylvania adopted a policy for "evolving research," such as ethnography, in which "research questions may only be clarified after a period of observation and where current findings drive the next steps in the study." Realizing that it was silly to ask researchers to request a new permission every time a finding led to a new question, the university allowed scholars with "documented discipline-appropriate education regarding human subject protection" a fair amount of leeway. Thus, while keeping social researchers under the jurisdiction of its Human Research Protection Program, the university excused them from cramming a qualitative, open-ended study into paperwork designed for experiments.[154]

All of these proposals would impose some oversight on research in the social sciences and humanities, but would do so with processes quite unlike those operating at most universities in the English-speaking world. They all envisioned transferring power from general boards (or "social-behavioral boards") dominated by medical and psychological researchers to scholars in the fields under review. What these ideas also have in common is that they were proposed in the early 1970s, but rejected by policy makers who cared little about the social sciences. In 1973 the Tuskegee Syphilis Study Ad Hoc Advisory Council argued that "publication of [IRB] decisions would permit their intensive study both inside and outside the medical profession and would be a first step toward the case-by-case development of policies governing human experimentation. We regard such a development, analogous to the experience of the common law, as the best hope for ultimately providing workable standards for the regulation of the human experimentation process."[155] But Congress rejected the idea of a national board, and the failure of the 1981 regulations to provide for an appeals process, or any coordination among IRBs, resulted in the highly localized, and therefore idiosyncratic, system of policy making and decision making that characterizes today's IRBs. Decentralization also emerged in the 1970s, only to be smothered by federal regulators who discouraged efforts "to put a sub-committee in every department."[156] Around the same time, regulators forbade the University of California, Berkeley from using the affidavit system that held such promise, killing that idea for more than thirty years.

In short, while reformers of the mid-2000s brought knowledge and creativity to their proposals for better ethical review, they were struggling against decades of policy decisions. Regulators had designed a system that allowed them to impose uniform structures of review on all manner of institutions and all manner of research, but that did not require the hard work of reconciling disagreements between boards, or even between decisions of a single board. In doing so, they left obstacles in the path of future reformers.

The horror stories of the 1990s and 2000s showed that Ithiel de Sola Pool had been largely correct in his predictions. Advocates of both accommodation and resistance could agree that the system as it existed in the late 2000s often—perhaps regularly—produced bad results. Of course, most scholars in the humanities and social sciences did their work without facing significant interference from IRBs, just as most scholars of the 1910s and 1950s had not been persecuted for suspected pacifist or communist beliefs. Rather, as in earlier assaults on scholarship, the burden of IRB abuse fell disproportionately on the most junior scholars, those with the most controversial research agendas, and those with the misfortune to work at the least adeptly administered universities.

In addressing the problems of IRB review, both accommodationists and resisters struggled against decades of history. Regulations and institutions still new and pliant in 1979 had become rigid by the late 1990s. A generation of researchers had been trained to accept IRB review as a standard step in developing a project, rather than questioning its necessity. Within the federal government and university administrations, key posts were occupied by people who could not imagine allowing social science research to proceed without prior review. And an entire industry had emerged to reassure them that committee review was the indispensible centerpiece to the ethical conduct of research. Reformers of all stripes should not have been surprised at the difficulty of shifting this structure. Changing it would mean challenging history.

CONCLUSION

Since 1966, American scholars have faced four different regimes of IRB review. The first, initiated in 1966 by the Public Health Service, applied mainly to medical and psychological researchers with PHS grants. By 1972, however, IRB oversight had begun to spread to social scientists, even those without direct federal funding. The second regime was negotiated between Congress, which passed the National Research Act, and the Department of Health, Education, and Welfare, whose 1974 regulations were less an attempt to put that act into practice than to allow DHEW to continue the path it had been taking without congressional interference. The third regime, initiated in 1981, relied heavily on the 1978 recommendations of the National Commission. But it also reflected demands from the President's Commission, from Congress, and from outside critics that the federal government should avoid imposing IRB oversight on research for which it was practically or ethically inappropriate. Though Charles McCarthy began chipping away at that compromise as soon as it was passed, it was not until around 1995 or 1998 that his successor, Gary Ellis, imposed a new, fourth regime that was based on the omnipresent threat of federal penalties. But though that regime—still in place today—is relatively new, it builds on trends present since the beginning. To understand, and remedy, problems faced by today's scholars, one must consider the achievements and the errors of the past.

History alone cannot prescribe future action, for the lessons of the past are always ambiguous. But human beings cannot reason without some ideas about past events, and when they lack real knowledge about the past, they are tempted to invent stories that suit them. In the absence of serious investigation into the origins of IRB review of the social sciences and humanities, policy makers and other advocates too easily accept reassuring, but misleading, accounts. Thus, while I have my own strongly held beliefs about the future of IRB review, I think I can serve the debate best by presenting some simple findings about the past.

The first finding is that the present system of IRB oversight is not based on empirical investigation of ethical abuses committed by social scientists. The IRB system in general is weak on empirical evidence, but at least the regulation of medical research was grounded in the work of Henry Beecher, Bernard Barber, Bradford Gray, the National Commission, the Advisory Committee on Human Radiation Experiments, and other commissions, as well as lengthy congressional hearings in the 1970s and again in the 1990s. No such official study of the rights and responsibilities of social scientists has ever been conducted, and what studies did exist on particular issues—like confidentiality—played virtually no role in shaping policy. Instead of educating themselves about the varied work and challenges of social science research, IRB advocates used the controversy surrounding Laud Humphreys as sufficient justification for restricting all social scientists. But Humphreys's story is less representative and more ambiguous than those advocates may realize. And forty years after Humphreys did his work, they may need a new demon.

This is not to say that social scientists do not make ethical mistakes. "I had been so wrapped up in my desire to obtain good data that I couldn't anticipate the consequences of my actions," wrote sociologist Sudhir Venkatesh, recalling how by exposing their finances he had made his informants vulnerable to extortion.[1] But such incidents are rare, and, as Gary Ellis put it, "we're talking about the numerator of a very large denominator."[2] If governments and universities are serious about protecting the freedom of inquiry, they may have to accept that one unethical project will go through in order that a thousand—or ten thousand—ethical projects can proceed without interference.

The second finding is that policy makers failed to explore alternative measures to prevent such abuses as do occur. It is by no means clear that even frequent wrongs by social scientists would merit prospective review of social research, given that IRBs have proven so inept at tailoring their recommendations to the variety of methods used and of topics explored by social scientists. That Venkatesh could not anticipate the consequences of his actions is no reason to think that a committee of experimental psychologists, education researchers, and laypeople could have done better. And while calls for more research are always welcome, the burden of proof should be on those who would restrict liberty. When they insist that "research is a privilege," IRB defenders excuse themselves from the hard work of showing IRBs to be safe and efficacious, and from investigating other options. The clearest example of this concerns the confidentiality of research data. As scholars have shown since the 1970s, the most serious threat to this confidentiality is not an unethical researcher, but a government subpoena. If governments were serious about protecting confidentiality, they would put less emphasis on ethics committees and instead pass laws shielding research data from subpoena.

The third finding is that medical and psychological researchers have been well represented on every official body that has set IRB policy, dating from the National Advisory Health Council meetings of 1965 through the current Secretary's Advisory Committee on Human Research Protections, as well as federal agency staffs and nonprofit groups like PRIM&R. In contrast, official bodies have included at most a token representation from the social sciences (often a lone medical sociologist or medical anthropologist), and even panels dedicated to nonbiomedical research have failed to distinguish important differences between experimental fields—like psychology and education—and more open-ended, qualitative work—such as ethnography and oral history. It is because the consent of the governed builds respect for law that Congress tapped eminent medical and psychological researchers as witnesses and commission members as it framed its system of ethical regulation of medical and psychological research. Scholars in the social sciences and the humanities have never been accorded comparable respect. It is no wonder that so many social researchers, frozen out of the process, regard its product with contempt.

The fourth finding is that the extension of IRB oversight over most social science research was largely unintentional, or at least so flawed that no one has been willing to take responsibility for it. University officials point to federal rules, while the authors of those rules claimed they were powerless to avoid the creep of regulation. Bradford Gray of the National Commission staff "felt like we were stuck in including social research within the boundaries of research involving human subjects."[3] David Kefauver of the NIMH worried that "if we deregulate or fail to regulate by topical sensitivity . . . we would have to deregulate biomedical research."[4] Gary Ellis of the OPRR tightened the regulation of social research in response to presidential concerns about radiation experiments. For such officials, the goal was to contain medical research, and the regulation of the social sciences was merely collateral damage. A possible exception to this pattern was Charles McCarthy's determined effort to beat back the challenge of Pattullo and Pool, but McCarthy's explanations of that effort are so varied and doubtful that one cannot say just what he was trying to do.

In contrast, health officials have repeatedly asserted their wish to deregulate most or all social research. In September 1966, Mordecai Gordon promised that the PHS would work to exclude anthropological and sociological field studies. In 1974, DHEW promised that "policies are also under consideration which will be particularly concerned . . . with the subject of social science research," but it never developed those policies. In 1981, HHS claimed its new policies would "exclude most social science research projects from the jurisdiction of the regulations." More recently, federal officials have blamed universities for failing to use the "flexibility" in the regulations.[5] By the standards set by regulators themselves, the routine review of social science by IRBs is a mistake. But no one has made a serious effort to correct that mistake.

A fifth finding is that while some scholars look back to what Robert Levine terms "the good old days" of IRB review of medical research before the 1998 OPRR crackdown, there has never been a golden age of IRB review of the social sciences.[6] The early years of IRB review were marked by abuses of power at major universities, leading to the outraged testimony collected by the National Commission and the President's Commission. The years 1981 through 1995 mark a period of peace, not because IRBs did a good job in reviewing social research, but because they did a good job of leaving social scientists alone—not insisting that they seek permission to do their work. When regulators again asserted authority over social research in the 1990s, the horror stories started up again. Those social scientists who call for freedom from IRB review are looking back to a regime of noninterference that once worked. Those who want to adapt IRB review to a broad range of social science and humanities work are calling for something not yet seen in forty years of policy making.

The final finding is that the creators of today's IRB system treated history as carelessly as they treated historians—and other scholars. They claimed to learn from the past, pointing out the wrongs of the Tuskegee Syphilis Study and the human radiation experiments, and noting the longevity of the IRB system as a reason to continue it. But they have dismissed the past when it suited them, such as when Gary Ellis's OPRR of the 1990s reinterpreted the Common Rule's exemption clauses to require some oversight of even exempt research, and when it decided that oral history was generalizable. Ellis made these decisions knowing, but not caring, that he was reading the regulations in ways that their authors would not recognize. Given the inconsistent treatment of the past, it might be best for regulators to start afresh. David Strauss, a member of the Secretary's Advisory Committee on Human Research Protections, put it nicely in 2008 when he lamented, "We really shouldn't be reviewing research that we don't feel needs to be reviewed simply because some folks 30 years ago . . . at the end of a long, hot day, you know, decided to use the word 'generalizable.'"[7] Strauss's recognition—that neither the *Belmont Report* nor the Common Rule were the work of gods—could point the way toward a more open-minded approach to today's problems.

We also need to place responsibility where it belongs. While the people most directly responsible for abusing social scientists are generally IRB administrators and members, ultimate responsibility lies with the federal government. Since the 1960s, federal regulators have been telling universities that unless they impose close oversight on their social scientists, they risk the loss of federal research funds that make up a significant share of university budgets. In some cases, as at Berkeley and the University of Colorado in the 1970s, federal officials played a direct role in shaping university policy. In other cases, the influence has been less direct, but equally important. While reformers have made some imaginative proposals for universities to work within the present structure of federal policy, the history of IRBs suggests that true, systemic reform will only occur if that structure itself changes.

Federal policy, in turn, exists at several levels: guidance, regulation, and statutory law. Possible avenues of reform could include the issuance of new guidance, a change in regulations, or litigation that would invalidate policies deemed unconstitutional or otherwise illegal. The problem with all of these steps is that they would leave the interpretation of regulations in the hands of officials whose main job is regulating medical research. So long as that is the case, real change is unlikely. For all their talk about potential conflicts of interests among researchers, federal officials have been silent about the conflict of interest that lies at the heart of their own work. An official hired to protect participants in medical experimentation is bound to focus on that life-or-death, multibillion-dollar task, even if it means ignoring equally important, but lower-profile, social science research. Whoever occupies the office, the pressures are the same.

My own hope, then, would be for Congress to relieve health regulators of the responsibility for overseeing the social sciences, a task they have mishandled for decades. Congress should amend the National Research Act to restrict its scope to the research discussed during the 1973 Senate hearings that still serve as the evidentiary record for that act. The wording of such a restriction can be drawn from Pattullo's formula of 1979: "There should be no requirement for prior review of research utilizing legally competent subjects if that research involves neither deceit, nor intrusion upon the subject's person, nor denial or withholding of accustomed or necessary resources."

Had such a restriction been included in the law in 1974, social scientists probably would not face significant IRB problems now. Amending the statute today would have less effect. Still in place would be state laws governing human subjects research, laws in other nations based on the U.S. example, and, most importantly, a whole industry of university administrators and consultants who believe that IRB review of the social sciences is necessary. But all of these branches of IRB power draw sustenance from the ill-drafted statute at their root. Cut out that root, and the tangled branches may yet wither.

When wise people, fairly selected, craft policies based on careful investigation and deliberation, their decisions deserve respect, whether the result is a constitution or a humble regulation. But when policy makers deny power to experts and to representatives of those most affected by the restrictions, when they ignore available evidence, when they rush regulations into print along with empty promises of future revision, and when they restrict freedom simply as an afterthought, their actions deserve little deference. IRB review of the social sciences and the humanities was founded on ignorance, haste, and disrespect. The more people understand the current system as a product of this history, the more they will see it as capable of change.

NOTES

ABBREVIATIONS

DAR	Donald A. Ritchie, private collection, Washington, D.C.
IdSP	Ithiel de Sola Pool Papers, MC 440, Institute Archives and Special Collections, MIT Libraries, Cambridge, Massachusetts
NC-GTU	National Commission for the Protection of Human Subjects of Biomedical and Behavioral Research Collection, 1974–78, Graduate Theological Union, Berkeley, California
NCPHS-GU	National Commission for the Protection of Human Subjects of Biomedical and Behavioral Research Collection, National Reference Center for Bioethics Literature, Kennedy Institute of Ethics, Georgetown University, Washington, D.C.
OD-NIH	OD Central Files, Office of the Director, National Institutes of Health, Bethesda, Maryland
PCSEP-GU	President's Commission for the Study of Ethical Problems in Medicine and Biomedical and Behavioral Research Records, National Reference Center for Bioethics Literature, Kennedy Institute of Ethics, Georgetown University, Washington, D.C.
RG 443	Central Files of the Office of the Director, National Institutes of Health, 1960–1982, National Archives II, Record Group 443, College Park, Maryland
UCB	Archives, University Library, University of Colorado at Boulder

Recordings and, in some cases, transcripts of interviews by the author have been deposited as the Institutional Review Board Oral History Project, Special Collections & Archives, Fenwick Library, George Mason University, Fairfax, Virginia.

INTRODUCTION

1. Sherryl Browne-Graves and Andrea Savage to Bernadette McCauley, 18 Nov. 2004, in the author's possession.

2. Bernadette McCauley to Frederick P. Shaffer, 15 Feb. 2005, in the author's possession.

3. Bernadette McCauley, "An IRB at Work: A Personal Experience," *Perspectives: Newsletter of the American Historical Association* (Feb. 2006); Patricia Cohen, "As Ethics Panels Expand Grip, No Field Is Off Limits," *New York Times*, 8 Feb. 2007; author's conversations with McCauley, July and Nov. 2008.

4. See chapter 8 for examples of research in these areas.

5. Scott Burris and Kathryn Moss, "U.S. Health Researchers Review Their Ethics Review Boards: A Qualitative Study," *Journal of Empirical Research on Human Research Ethics* 2 (June 2006): 39–58; Simon N. Whitney, Kirsten Alcser, Carl E. Schneider, Laurence B. McCullough, Amy L. McGuire, and Robert J. Volk, "Principal Investigator Views of the IRB System," *International Journal of Medical Sciences* 5 (April 2008): 68–72.

6. Personal statement of Carl E. Schneider, J.D., in President's Council on Bioethics, *The Changing Moral Focus of Newborn Screening: An Ethical Analysis by the President's Council on Bioethics* (Washington, DC: President's Council on Bioethics, 2008).

7. For example, several law professors contributed to a special issue of the *Northwestern University Law Review* in 2007.

8. Rena Lederman, "The Perils of Working at Home: IRB 'Mission Creep' as Context and Content for an Ethnography of Disciplinary Knowledges," *American Ethnologist* 33 (Nov. 2006): 482–491.

9. Debbie S. Dougherty and Michael W. Kramer, "A Rationale for Scholarly Examination of Institutional Review Boards: A Case Study," *Journal of Applied Communication Research* 33 (Aug. 2005): 183–188; also see the other articles in that issue.

10. Jeffrey M. Cohen, Elizabeth Bankert, and Jeffrey A. Cooper, "History and Ethics," CITI Course in the Protection of Human Research Subjects, www.citiprogram.org [accessed 30 Oct. 2006; available only to affiliates of participating organizations].

11. Robert S. Broadhead, "Human Rights and Human Subjects: Ethics and Strategies in Social Science Research," *Sociological Inquiry* 54 (Apr. 1984): 107. Broadhead cites two articles about controversial research, neither of which shows that such controversy led to IRBs' powers.

12. Laura Stark, "Victims in Our Own Minds? IRBs in Myth and Practice," *Law & Society Review* 41 (Dec. 2007): 779.

13. Eleanor Singer and Felice J. Levine, "Protection of Human Subjects of Research: Recent Developments and Future Prospects for the Social Sciences," *Public Opinion Quarterly* 67 (Spring 2003): 149.

14. Murray L. Wax, "Human Rights and Human Subjects: Strategies in Social Science Research," *Sociological Inquiry* 55 (Oct. 1985): 423.

15. Lederman, "Perils of Working at Home," 486.

16. Robert L. Kerr, "Unconstitutional Review Board? Considering a First Amendment Challenge to IRB Regulation of Journalistic Research Methods," *Communication Law and Policy* 11 (Summer 2006): 412. The quotation refers to the Tuskegee Syphilis Study, which will be discussed in chapter two.

17. C. K. Gunsalus, Edward M. Bruner, Nicholas C. Burbules, Leon Dash, Matthew Finkin, Joseph P. Goldberg, William T. Greenough, Gregory A. Miller, Michael G. Pratt, Masumi Iriye, and Deb Aronson, "The Illinois White Paper: Improving the System for

Protecting Human Subjects: Counteracting IRB 'Mission Creep,'" *Qualitative Inquiry* 13 (July 2007): 618.

18. Constance F. Citro, Daniel R. Ilgen, and Cora B. Marrett, eds., *Protecting Participants and Facilitating Social and Behavioral Sciences Research* (Washington, DC: National Academies Press, 2003), 59.

19. National Commission, transcript, meeting #41, Apr. 1978, II-13, NCPHS-GU. Commission transcripts can be found in boxes 24–33, NCPHS-GU. In 2007, Jonsen described today's human subjects regulations as an example of ethical imperialism (Albert Jonsen, interview by the author, San Francisco, 24 Oct. 2007), and he kindly granted me permission to use the phrase as the title of this book.

20. John Robert Seeley, *The Expansion of England: Two Courses of Lectures* (Boston: Little, Brown, 1922), 10.

CHAPTER 1: ETHICS AND COMMITTEES

1. Dorothy Ross, *The Origins of American Social Science* (New York: Cambridge University Press, 1991), 3–8.

2. Peter T. Manicas, "The Social Science Disciplines: The American Model," in Peter Wagner, Bjèorn Wittrock, and Richard Whitley, eds., *Discourses On Society: The Shaping of the Social Science Disciplines* (Boston: Kluwer Academic Publishers, 1991), 46.

3. Ross, *Origins of American Social Science*, 63.

4. Daniel G. Brinton, "The Aims of Anthropology," *Science* 2 (30 Aug. 1895): 242.

5. American Association for the Advancement of Science, "AAAS Sections," www.aaas.org [accessed 8 Feb. 2008].

6. George W. Stocking Jr., "Franz Boas and the Founding of the American Anthropological Association," *American Anthropologist* 62 (Feb. 1960): 1–17.

7. Ross, *Origins of American Social Science*, chapter 8.

8. Manicas, "Social Science Disciplines," 64.

9. Edward H. Spicer, "Beyond Analysis and Explanation? Notes on the Life and Times of the Society for Applied Anthropology," *Human Organization* 35 (Winter 1976): 336.

10. Roger Smith, *The Norton History of the Human Sciences* (New York: W. W. Norton, 1997), 520–528.

11. Thomas M. Camfield, "The Professionalization of American Psychology, 1870–1917," *Journal of the History of the Behavioral Sciences* 9 (Jan. 1973): 69–73.

12. Debora Hammond and Jennifer Wilby, "The Life and Work of James Grier Miller," *Systems Research and Behavioral Science* 23 (May/June 2006): 430–432.

13. Arthur W. Macmahon, "Review of *A Report on the Behavioral Sciences at the University of Chicago*," *American Political Science Review* 49 (Sept. 1955): 859.

14. Macmahon, "Review of *A Report*," 862.

15. Charles D. Bolton, "Is Sociology a Behavioral Science?" *Pacific Sociological Review* 6 (Spring 1963): 3–9.

16. Carolyn Fluehr-Lobban, "Ethics and Professionalism: A Review of Issues and Principles within Anthropology," in Carolyn Fluehr-Lobban, ed., *Ethics and the Profession*

of Anthropology: Dialogue for Ethically Conscious Practice (Philadelphia: University of Pennsylvania Press, 1991), 17–20; David H. Price, *Anthropological Intelligence: The Deployment and Neglect of American Anthropology in the Second World War* (Durham, NC: Duke University Press, 2008), 266–273.

17. David Price, "Interlopers and Invited Guests: On Anthropology's Witting and Unwitting Links to Intelligence Agencies," *Anthropology Today* 18 (Dec. 2002): 16–20.

18. Irving Louis Horowitz, "The Rise and Fall of Project Camelot," in Irving Louis Horowitz, ed., *The Rise and Fall of Project Camelot: Studies in the Relationship between Social Science and Practical Politics*, rev. ed. (Cambridge, MA: MIT Press, 1974), 5–13.

19. Joseph G. Jorgensen and Eric R. Wolf, "A Special Supplement: Anthropology on the Warpath in Thailand," *New York Review of Books* (19 Nov. 1970); George M. Foster, Peter Hinton, A. J. F. Köbben and reply by Joseph G. Jorgensen, Eric R. Wolf, "Anthropology on the Warpath: An Exchange," *New York Review of Books* (8 Apr. 1971).

20. Seymour J. Deitchman, *The Best-Laid Schemes: A Tale of Social Research and Bureaucracy* (Cambridge, MA: MIT Press, 1976), 421–422.

21. Joy Rohde, "Gray Matters: Social Scientists, Military Patronage, and Democracy in the Cold War," *Journal of American History* 96 (June 2009): 99–122.

22. Herbert C. Kelman, *A Time to Speak: On Human Values and Social Research* (San Francisco: Jossey-Bass, 1968), chapter 8; Thomas Blass, *The Man Who Shocked the World: The Life and Legacy of Stanley Milgram* (New York: Basic Books, 2004), chapter 7.

23. Kai T. Erikson, "A Comment on Disguised Observation in Sociology," *Social Problems* 14 (Spring 1967): 367.

24. Howard S. Becker, "Problems in the Publication of Field Studies," in Arthur J. Vidich, Joseph Bensman, and Maurice R. Stein, eds., *Reflections on Community Studies* (New York: John Wiley & Sons, 1964), 268; Myron Glazer, *The Research Adventure: Promise and Problems of Field Work* (New York: Random House, 1972), 30, 103–106.

25. Erikson, "Comment on Disguised Observation," 366; Norman K. Denzin, "On the Ethics of Disguised Observation," *Social Problems* 15 (Spring 1968): 502–504; Kai T. Erikson, "A Reply to Denzin," *Social Problems* 15 (Spring 1968): 505–506.

26. Margaret Mead, "Research with Human Beings: A Model Derived from Anthropological Field Practice," *Daedalus* 98 (Spring 1969): 363.

27. J. E. Hulett Jr., "Interviewing in Social Research: Basic Problems of the First Field Trip," *Social Forces* 16 (Mar. 1938): 365.

28. The creation of pseudonyms for whole communities began in the 1920s to "lend an aura of typicality" rather than to protect informants. Stephan Thernstrom, *Poverty and Progress: Social Mobility in a Nineteenth Century City* (Cambridge, MA: Harvard University Press, 1964), cited in Robert Roy Reed and Jay Szklut, "The Anonymous Community: Queries and Comments," *American Anthropologist* 90 (Sept. 1988): 689.

29. Arthur J. Vidich and Joseph Bensman, "'Freedom and Responsibility in Research': Comments," *Human Organization* 17 (Winter 1958–59), reprinted in Arthur J. Vidich and Joseph Bensman, *Small Town in Mass Society: Class, Power, and Religion in a Rural Community*, rev. ed. (Urbana: University of Illinois Press, 2000), 405.

30. Vidich and Bensman, *Small Town in Mass Society*, chapter 14.

31. Harold Orlans, "Ethical Problems and Values in Anthropological Research," in House Committee on Government Operations, Research and Technical Programs Subcommittee, *The Use of Social Research in Federal Domestic Programs: Part 4, Current Issues in the Administration of Federal Social Research*, 90th Cong., 1st sess., 1967, 362.

32. Ralph L. Beals, *Politics of Social Research: An Inquiry into the Ethics and Responsibilities of Social Scientists* (Chicago: Aldine, 1969), 35.

33. Lee Rainwater and David J. Pittman, "Ethical Problems in Studying a Politically Sensitive and Deviant Community," *Social Problems* 14 (Spring 1967): 364–365.

34. James D. Carroll and Charles R. Knerr Jr., "The APSA Confidentiality in Social Science Research Project: A Final Report," *PS: Political Science and Politics* 9 (Autumn 1976): 418.

35. Joseph H. Fichter and William L. Kolb, "Ethical Limitations on Sociological Reporting," *American Sociological Review* 18 (Oct. 1953): 544–550.

36. Brian M. du Toit, "Ethics, Informed Consent, and Fieldwork," *Journal of Anthropological Research* 36 (Autumn 1980): 277.

37. Mead, "Research with Human Beings," 361.

38. Gresham M. Sykes, "Feeling Our Way: A Report on a Conference on Ethical Issues in the Social Sciences," *American Behavioral Scientist* 10 (June 1967): 9.

39. Fichter and Kolb, "Ethical Limitations," 549.

40. William F. Whyte, "Freedom and Responsibility in Research: The 'Springdale' Case," *Human Organization* 17 (1958–59), reprinted in Vidich and Bensman, *Small Town in Mass Society*, 401.

41. Rainwater and Pittman, "Ethical Problems," 361, 366.

42. Mead, "Research with Human Beings," 366.

43. Becker, "Problems in the Publication of Field Studies," 271.

44. "Minutes of the 1963 Annual Business Meeting," *Human Organization* 22 (Winter 1964): 313; "Statement on Ethics of the Society for Applied Anthropology," *Human Organization* 22 (Winter 1964): 237.

45. Society for Applied Anthropology, "Statement on Professional and Ethical Responsibilities," 13 Mar. 1974, box 8, meeting #18, tab 14, NCPHS-GU.

46. Fluehr-Lobban, "Ethics and Professionalism," 25.

47. American Anthropological Association, "Statement on Problems of Anthropological Research and Ethics," Mar. 1967, Codes of Ethics Collection, Center for the Study of Ethics in the Professions, Illinois Institute of Technology, Chicago, Illinois.

48. "Full Report of Ethics Committee Presented by Board," *Newsletter of the American Anthropological Association* (Nov. 1970): 15.

49. Foster et al., "Anthropology on the Warpath."

50. Glazer, *The Research Adventure*, 96.

51. "Toward a Code of Ethics for Sociologists," *American Sociologist* (Nov. 1968): 316–318.

52. Paul Davidson Reynolds, *Value Dilemmas Associated with the Development and Application of Social Science* (Paris: UNESCO, 1975), II-82.

53. American Political Science Association, "Final Report of the American Political Science Association Committee on Professional Standards and Responsibilities: Ethical

Problems of Academic Political Scientists," *PS: Political Science and Politics* 1 (Summer 1968): 3–29.

54. William Manchester, *Controversy and Other Essays in Journalism, 1950–1975* (Boston: Little, Brown, 1976), 3–76.

55. Willa Baum, "A Code of Ethics for Oral Historians?" *Oral History Association Newsletter* (Apr. 1968): 3.

56. "Oral History Association Adopts Statement about Goals and Guidelines during Nebraska Colloquium," *Oral History Association Newsletter* (Jan. 1969): 4.

57. "Proposed Revision of OHA Goals and Guidelines," *Oral History Association Newsletter* (Fall 1975): 4; "Oral History Evaluation Guidelines: The Wingspread Conference," *Oral History Review* 8 (issue 1, 1980): 8.

58. American Psychological Association, *Ethical Principles in the Conduct of Research with Human Participants* (Washington, DC: American Psychological Association, 1973), 3.

59. Nicholas Hobbs, "The Development of a Code of Ethical Standards for Psychology," *American Psychologist* 3 (Mar. 1948): 82–83.

60. American Psychological Association, *Ethical Principles*, 5–6.

61. Reynolds, *Value Dilemmas*, II-82.

62. Howard S. Becker, "Against the Code of Ethics," *American Sociological Review* 29 (June 1964): 409–410; Eliot Freidson, "Against the Code of Ethics," *American Sociological Review* 29 (June 1964): 410; Köbben in Foster et al., "Anthropology on the Warpath."

63. Becker, "Against the Code of Ethics."

64. John F. Galliher, Wayne H. Brekhus, and David P. Keys, *Laud Humphreys: Prophet of Homosexuality and Sociology* (Madison: University of Wisconsin Press, 2004), 27.

65. Laud Humphreys, *Tearoom Trade: Impersonal Sex in Public Places*, enlarged ed. (Chicago: Aldine, 1975), xviii.

66. Humphreys, *Tearoom Trade* (enl. ed.), 25.

67. Galliher et al., *Laud Humphreys*, 29.

68. Laud Humphreys, "Social Science: Ethics of Research," *Science* 207 (15 Feb. 1980): 712.

69. Humphreys, *Tearoom Trade* (enl. ed.), 24 n. 16, 30–42, 104, 169.

70. Nicholas von Hoffman, "Sociological Snoopers," *Washington Post*, 30 Jan. 1970.

71. Laud Humphreys, *Tearoom Trade: Impersonal Sex in Public Places* [1st ed.] (Chicago: Aldine, 1970).

72. Donald P. Warwick, "*Tearoom Trade*: Means and Ends in Social Research," *Hastings Center Studies* 1 (issue 1, 1973): 31–32, 37.

73. Irving Louis Horowitz and Lee Rainwater, "On Journalistic Moralizers," *Trans-Action* 7 (May 1970), reprinted in Humphreys, *Tearoom Trade* (enl. ed.), 187.

74. Julia S. Brown and Brian G. Gilmartin, "Sociology Today: Lacunae, Emphases, and Surfeits," *American Sociologist* 4 (Nov. 1969): 288.

75. Richard A Shweder, "Tuskegee Re-Examined," *spiked-essays*, 8 Jan. 2004, www.spiked-online.com/Articles/0000000CA34A.htm [accessed 2 May 2008].

76. Galliher et al., *Laud Humphreys*, 27; Glazer, *The Research Adventure*, 114.

77. Humphreys, *Tearoom Trade* (enl. ed.), 227.

78. Humphreys, "Social Science," 714.

79. Galliher et al., *Laud Humphreys*, 72–73.

80. Warwick, "*Tearoom Trade*: Means and Ends," 29.

81. Glazer, *The Research Adventure*, 112, 116.

82. Humphreys, *Tearoom Trade* (enl. ed.), 230.

83. Victoria A. Harden, "A Short History of the National Institutes of Health," http://history.nih.gov/exhibits/history/index.html [accessed 3 Mar. 2008]; Albert R. Jonsen, *The Birth of Bioethics* (New York: Oxford University Press, 1998), 142.

84. Robert N. Proctor, "Nazi Doctors, Racial Medicine, and Human Experimentation," in George J. Annas and Michael A. Grodin, eds., *The Nazi Doctors and the Nuremberg Code: Human Rights in Human Experimentation* (New York: Oxford University Press, 1992).

85. David M. Oshinsky, *Polio: An American Story* (New York: Oxford University Press, 2005), 226, 238.

86. David J. Rothman and Sheila M. Rothman, *The Willowbrook Wars: Bringing the Mentally Disabled into the Community* (1984; reprint, New Brunswick, NJ: Aldine Transaction, 2005), 260–266; Walter Sullivan, "Project on Hepatitis Research Is Now Praised by State Critic," *New York Times*, 24 Mar. 1971.

87. Elinor Langer, "Human Experimentation: Cancer Studies at Sloan-Kettering Stir Public Debate on Medical Ethics," *Science* 143 (7 Feb. 1964): 552.

88. James A. Shannon, interview by Mark Frankel, New York City, 13 May 1971, recording in the author's possession.

89. House Committee on Government Operations, *Use of Social Research, Part 4*, 217.

90. William J. Curran, "Governmental Regulation of the Use of Human Subjects in Medical Research: The Approach of Two Federal Agencies," *Daedalus* 98 (Spring 1969): 575.

91. PHS, "PPO #129: Clinical Investigations Using Human Subjects," 8 Feb. 1966, Res 3-1, Human Subjects Policy & Regulations 1965–67, RG 443.

92. "PPO #129, revised policy, 1 July 1966," in Senate Committee on Labor and Human Resources, *National Advisory Commission on Health Science and Society*, 92nd Cong., 1st sess., 1971, 66.

93. James A. Haggarty, "Applications Involving Research on Human Subjects," 28 Feb. 1966, Res 3-1, Human Subjects Policy & Regulations 1965–67, RG 443.

94. NIH Study Committee, *Biomedical Science and Its Administration* (Washington, DC: GPO, 1965), 130–131. Psychology seems to have dominated. In fiscal year 1972, for example, the NIH awarded fifty-eight research training grants in psychology, thirteen in sociology, and six in anthropology. Senate Committee on Labor and Human Resources, *National Research Service Award Act*, S. Report 93-381, 93rd Cong., 1st sess., 1973, 10.

95. Dael Wolfle to Patricia Harris, 11 Dec. 1980, box 25, Human Subjects Corresp. 1981 1/2, IdSP.

96. Shannon, interview by Frankel.

97. Ron Felber, *The Privacy War: One Congressman, J. Edgar Hoover, and the Fight for the Fourth Amendment* (Montvale, NJ: Croce, 2003), 46–48; Cornelius E. Gallagher, telephone interview by the author, 6 June 2008.

98. John D. Morriss, "House Unit Opens Polygraph Study," *New York Times*, 8 Apr. 1964.

99. Gallagher, telephone interview by the author.

100. Cornelius E. Gallagher, "Why House Hearings on Invasion of Privacy," *American Psychologist* (Nov. 1965), reprinted in House Committee on Government Operations, *Special Inquiry on Invasion of Privacy*, 89th Cong., 1st sess., 1966, 397–399.

101. House Committee on Government Operations, *Special Inquiry on Invasion of Privacy*, 5, 192.

102. House Committee on Government Operations, *Special Inquiry on Invasion of Privacy*, 38, 91, 136.

103. House Committee on Government Operations, *Special Inquiry on Invasion of Privacy*, 278–281.

104. House Committee on Government Operations, *Special Inquiry on Invasion of Privacy*, 301.

105. House Committee on Government Operations, *Special Inquiry on Invasion of Privacy*, 19–22.

106. House Committee on Government Operations, *Special Inquiry on Invasion of Privacy*, 355.

107. House Committee on Government Operations, *Special Inquiry on Invasion of Privacy*, 295.

108. Cornelius Gallagher to Luther L. Terry, Surgeon General, 13 Sept. 1965, Res 3-1, Human Subjects Policy & Regulations 1965–67, RG 443.

109. Morris S. Ogul, *Congress Oversees the Bureaucracy: Studies in Legislative Supervision* (Pittsburgh: University of Pittsburgh Press, 1976), 93.

110. Gallagher, telephone interview by the author.

111. Gallagher, telephone interview by the author.

112. Philip R. Lee to Cornelius E. Gallagher, 22 Nov. 1965, Res 3-1, Human Subjects Policy & Regulations 1965–67, RG 443.

113. PHS, "PPO #129," RG 443.

114. Haggarty, "Applications Involving Research on Human Subjects," RG 443.

CHAPTER 2: THE SPREAD OF INSTITUTIONAL REVIEW

1. Mordecai Gordon, interview by Mark S. Frankel, Bethesda, Maryland, 27 Apr. 1971, recording in the author's possession.

2. Sykes, "Feeling Our Way," 11.

3. Mead, "Research with Human Beings," 361.

4. American Sociological Association, "Resolution No. 9," 1 Sept. 1966, in House Committee on Government Operations, *Use of Social Research, Part 4*, 256–257.

5. "Minutes," *Social Problems* 14 (Winter 1967): 347.

6. Ernest M. Allen, "PPO #129, Revised, Supplement #2," 16 Dec. 1966, Res 3-1, Human Subjects Policy & Regulations 1965–67, RG 443.

7. Gordon, interview by Frankel.

8. Mordecai Gordon to Ralph S. Halford, 7 Sept. 1966, in House Committee on Government Operations, *Use of Social Research, Part 4*, 255.

9. PHS, "Investigations Involving Human Subjects, Including Clinical Research: Requirements for Review to Insure the Rights and Welfare of Individuals; *Clarification*," 12 Dec. 1966, Res 3-1, Human Subjects Policy & Regulations 1965–67, RG 443.

10. House Committee on Government Operations, *Use of Social Research, Part 4*, 220.

11. House Committee on Government Operations, Research and Technical Programs Subcommittee, *The Use of Social Research in Federal Domestic Programs: Part 3, The Relation of Private Social Scientists to Federal Programs on National Social Problems*, 90th Cong., 1st sess., 1967, 104.

12. House Committee on Government Operations, *Use of Social Research, Part 3*, 75.

13. House Committee on Government Operations, *Use of Social Research, Part 3*, 12.

14. House Committee on Government Operations, *Use of Social Research, Part 4*, 15.

15. "Privacy and Behavioral Research," *Science* 155 (3 Feb. 1967): 535–538.

16. Richard Mandel, *A Half Century of Peer Review, 1946–1996* (Bethesda, MD: National Institutes of Health, 1996), 126.

17. Mark Frankel, "Public Policymaking for Biomedical Research: The Case of Human Experimentation," PhD diss., George Washington University, 1976, 167.

18. Roger O. Egeborg, Assistant Secretary for Health and Scientific Affairs, to Director, NIH, 9 Feb. 1970, Res 3-1, Human Subjects Policy & Regulations 1966–72, RG 443.

19. "Origins of the DHEW Policy on Protection of Human Subjects," in Senate Committee on Labor and Human Resources, *National Advisory Commission on Health Science and Society*, 92nd Cong., 1st sess., 1971, 1–3.

20. Mark E. Conner, "Summary Minutes: DHEW Policy on Protection of Human Subjects, Meeting at Massachusetts General Hospital, April 30, 1971," Res 3-1, Human Subjects Policy & Regulations 1966–72, RG 443.

21. "Origins of the DHEW Policy on Protection of Human Subjects," 4.

22. DHEW, *The Institutional Guide to DHEW Policy on Protection of Human Subjects* (Washington, DC: GPO, 1971), 2.

23. DHEW, *Institutional Guide*, 2–3, 23–24.

24. James H. Jones, *Bad Blood: The Tuskegee Syphilis Experiment* (1981; reprint, New York: Free Press, 1993), 1.

25. Jean Heller, "Syphilis Victims in U.S. Study Went Untreated for 40 Years," *New York Times*, 26 July 1972.

26. *A Bill to Establish within the Executive Branch an Independent Board to Establish Guidelines for Experiments Involving Human Beings*, S. 934, 93rd Cong., 1st sess., 1973.

27. Jacob K. Javits, remarks, *Congressional Record* 119 (15 Feb. 1973): S 4235.

28. Senate Committee on Labor and Human Resources, *National Research Service Award Act*, 23.

29. House Committee on Interstate and Foreign Commerce, *Biomedical Research Ethics and the Protection of Human Research Subjects*, 93rd Cong., 1st sess., 1973, 240.

30. Senate Committee on Labor and Public Welfare, *Quality of Health Care: Human Experimentation, 1973, Part 3*, 93rd Cong., 1st sess., 1973, 1052.

31. Emanuel M. Kay, "Legislative History of Title II—Protection of Human Subjects of Biomedical and Behavioral Research of the *National Research Act*, P. L. 93-348," 1974, 12–13, box 1, folder 10, NC-GTU.

32. House Committee on Interstate and Foreign Commerce, *Biomedical Research Ethics*, 92.

33. Senate Committee on Labor and Human Resources, *National Research Service Award Act*, 14, 15, 24.

34. Senate Committee on the Judiciary, *Individual Rights and the Federal Role in Behavior Modification*, 93rd Cong., 2nd sess., 1974, 21.

35. Sharland Trotter, "Strict Regulations Proposed for Human Experimentation," *APA Monitor* 5 (Feb. 1974): 8.

36. Frankel, "Public Policymaking for Biomedical Research," 187, 298; Richard B. Stephenson to Robert P. Akers, 4 Oct. 1972, Res 3-1-A, Study Group Review of Policies 1973–75, RG 443.

37. Charles R. McCarthy, interview by Patricia C. El-Hinnawy, 22 July 2004, Oral History of the *Belmont Report* and the National Commission for the Protection of Human Subjects of Biomedical and Behavioral Research, www.hhs.gov/ohrp/belmontArchive.html/ [accessed 23 Dec. 2008]; Charles McCarthy to Study Group for Review of Policies on Protection of Human Subjects in Biomedical Research, 3 May 1973, Res 3-1, Human Subjects Policy & Regulations 1973–82, RG 443.

38. Frankel, "Public Policymaking for Biomedical Research," 187, 298.

39. Frankel, "Public Policymaking for Biomedical Research," 202.

40. Thomas J. Kennedy Jr. to Acting Director, NIH, 23 Mar. 1973, Res 3-1, Human Subjects Policy & Regulations 1973–82, RG 443.

41. Thomas J. Kennedy Jr. to Assistant Secretary for Health, 7 Sept. 1973, Res 3-1, Human Subjects Policy & Regulations 1973–82, RG 443.

42. D. T. Chalkley, draft letter to the editor of the *Christian Century*, 5 Apr. 1974, Res 3-4, National Comm. for the Protection of Human Sub., folder #1, 1974–1975, RG 443.

43. D. T. Chalkley to Mr. Secretary/Director/Commissioner, draft, 15 Oct. 1973, Res 3-1-B, Proposed Policy Protections Human Subjects 1973, RG 443.

44. Jerry W. Combs Jr., Ph.D., Chief, Behavioral Sciences Branch, Center for Population Research, to Dr. William Sadler, Acting Director, CPR, 25 Oct. 1973, Res 3-6-B, Proposed Policy Protection Human Subjects 1973, RG 443.

45. William A. Morrill to Bertram Brown, 22 Aug. 1973, Res 3-1-A, Study Group Review of Policies 1973–75, RG 443.

46. Director, NIH, to Assistant Secretary for Health, Designate, 11 Apr. 1977, Res 3-1, Human Subjects Policy & Regulations 1973–82, RG 443.

47. DHEW, "Protection of Human Subjects: Proposed Policy," *Federal Register* 38 (9 Oct. 1973): 27882.

48. DHEW, "Protection of Human Subjects," *Federal Register* 39 (30 May 1974): 18914, 18917.

49. Ronald Lamont-Havers to Robert S. Stone, 13 May 1974, Res 3-1-b-1, Interdept. Work. Group Uniform Fed. Pol., RG 443.

50. Thomas S. McFee to the Secretary, 5 June 1974, Res 3-1-B, Proposed Policy Protection Human Subjects 1974, RG 443.

51. McCarthy to Study Group for Review of Policies on Protection of Human Subjects in Biomedical Research, RG 443; David F. Kefauver, cited by Frankel, "Public Policymaking for Biomedical Research," 275, 320 n. 406.

52. Charles R. McCarthy, "Reflections on the Organizational Locus of the Office for Protection from Research Risks," in NBAC, *Commissioned Papers and Staff Analysis*, vol. 2 of *Ethical and Policy Issues in Research Involving Human Participants* (Bethesda, MD: National Bioethics Advisory Commission, 2001), H-8, http://bioethics.georgetown.edu/nbac/ [accessed 21 Oct. 2009].

53. Richard A. Tropp, "A Regulatory Perspective on Social Science Research," in Tom L. Beauchamp, Ruth R. Faden, R. Jay Wallace Jr., and LeRoy Walters, eds., *Ethical Issues in Social Science Research* (Baltimore: Johns Hopkins University Press, 1982), 39.

54. *National Research Act*, Public Law 93-348, 88 Stat. 342.

55. DHEW, "Protection of Human Subjects," 18917; "Mechanisms for Applying Ethical Principles to the Conduct of Research Involving Human Subjects: The Institutional Review Board," preliminary draft, 5 Nov. 1976, 7, box 11, meeting #24, tabs 5–6, NCPHS-GU.

56. *National Research Act*, § 212.

57. D. T. Chalkley to Mary F. Berry, 25 Aug. 1976, box 219, folder 3, Chancellor's papers, UCB.

58. Kingman Brewster, "Coercive Power of the Federal Purse," *Science* 188 (11 Apr. 1975): 105.

59. National Commission, transcript, meeting #1, Dec. 1974, 87.

60. *National Research Act*, § 202.

61. DHEW, "Protection of Human Subjects," 18914.

62. R. W. Lamont-Havers, "Discussion Paper re. DHEW Administration of Ethical Issues Relating to Human Research Subjects," draft, 3 Apr. 1974, Res 3-1-b-1, Interdept. Work. Group Uniform Fed. Pol., RG 443.

63. Ronald Lamont-Havers to Deputy Director, NIH, 5 Sept. 1974, Res 3-1-B, Proposed Policy Protection Human Subjects 1974, RG 443.

64. Richard J. Riseberg, Legal Advisor, NIH, Public Health Division, to Ronald Lamont-Havers, 29 Oct. 1974, Res 3-1-B, Proposed Policy Protection Human Subjects 1974, RG 443.

65. DHEW, "Protection of Human Subjects: Technical Amendments," *Federal Register* 40 (13 Mar. 1975): 11854–11858.

66. Eugene A. Confrey, "Status of Action to Revise Policy Statement on the Use of Human Subjects in PHS Sponsored Activities, and to Improve the Implementation of These Policies," 18 June 1968, Res 3-1, Human Subjects Policy & Regulations 1966–72, RG 443.

67. Stephen P. Hatchett, "Status Report of Experience with PPO #129," 31 May 1968, Res 3-1, Human Subjects Policy & Regulations 1966–72, RG 443.

68. Galliher et al., *Laud Humphreys*, 27.

69. Mark Conner, interview by Mark S. Frankel, 16 Feb. 1971, copy in the author's possession.

70. James Welsh, "Protecting Research Subjects: A New HEW Policy," *Educational Researcher* (Mar. 1972): 12.

71. Howard Higman, presentation, National Association of State Universities and Land-Grant Colleges, Seminar on Safeguards for Human Subjects in Research, 16 Jan. 1978, box 25, Human Subjects Recent Info., Part II, 11/79, 1/3, IdSP.

72. Fred Greenstein to Ithiel de Sola Pool, 23 June 1979, box 26, Human Subjects Thank You Letters 2/2, IdSP.

73. McCarthy, "Reflections on the Organizational Locus," H-6. McCarthy suggests that the change took place in 1972; in fact, it took effect on 27 October 1974.

74. John A. Robertson, "The Law of Institutional Review Boards," *UCLA Law Review* 26 (Feb. 1979): 497.

75. National Commission, transcript, meeting #24, Nov. 1976, 174, NCPHS-GU.

76. "Policy and Procedure of the University of California, Berkeley Campus, Governing the Protection of Human Subjects," 1 June 1972, box 8, folder 29, NC-GTU.

77. Herbert P. Phillips, "DHEW Regulations Governing the Protection of Human Subjects and Non-DHEW Research: A Berkeley View," box 3, meeting #10, tabs 13–14, NCPHS-GU.

78. Edward Shils, "Muting the Social Sciences at Berkeley," *Minerva* 11 (July 1973): 292–293; Allan Sindler and Phillip Johnson to Bernard Diamond, 26 Apr. 1973, box 8, folder 29, NC-GTU.

79. Herbert Phillips, interview by the author, Oakland, California, 23 Oct. 2007.

80. Herbert Phillips to members, Committee for the Protection of Human Subjects and the Academic Freedom Committee, 2 May 1973, box 8, folder 29, NC-GTU.

81. Herbert Phillips to Bernard L. Diamond, 2 Apr. 1973, box 8, folder 29, NC-GTU.

82. Herbert Phillips to Dominick R. Vetri, 10 June 1974, box 8, folder 29, NC-GTU.

83. Eugene J. Millstein, "The DHEW Requirements for the Protection of Human Subjects: Analysis and Impact at the University of California, Berkeley," July 1974, box 26, Human Subjects Thank You Letters 1/2, IdSP.

84. Phillips, interview by the author.

85. Herbert Phillips to Karen Lebacqz and David Louisell, 14 Mar. 1975, box 8, folder 29, NC-GTU.

86. Millstein, "DHEW Requirements," IdSP.

87. Phillips, "DHEW Regulations," NCPHS-GU.

88. Phillips, interview by the author.

89. Phillips, "DHEW Regulations," NCPHS-GU.

90. Louis A Wienckowski to John L. Horn, 9 Dec. 1974, box 8, folder 29, NC-GTU.

91. D. T. Chalkley to Benson Schaeffer, 20 Feb. 1975, box 8, folder 29, NC-GTU.

92. Phillips to Lebacqz and Louisell, 14 Mar. 1975, NC-GTU.

93. Phillips, interview by the author.

94. Jane Roberts to D. T. Chalkley, 27 Aug. 1976, box 219, folder 3, Chancellor's papers, UCB.

95. DHEW, "Protection of Human Subjects: Technical Amendments," 11854.

96. University of Colorado, "Institutional General Assurance," 19 Feb. 1976, box 5, folder 22, NC-GTU.

97. University of Colorado Human Research Committee, "Review of All Human Research at the University of Colorado," 9 Apr. 1976, box 5, folder 22, NC-GTU.

98. Edward Rose to Mary Berry, 16 July 1976, box 5, folder 22, NC-GTU.

99. Roberts to Chalkley, 27 Aug. 1976, UCB.

100. Edward Rose to David Mathews, 12 July 1976, box 5, folder 22, NC-GTU.

101. William Hodges to Department of Sociology, 17 Aug. 1976; Howard Higman to Mary Berry, 25 Aug. 1976; both box 219, folder 3, Chancellor's papers, UCB.

102. Howard Higman to William Hodges, 25 Aug. 1976; William Hodges to Edward Rose, 13 Sept. 1976; both box 219, folder 3, Chancellor's papers, UCB.

103. Edward Rose to Jonathan B. Chase, 30 July 1976; Edward Rose to Mary Berry, 16 July 1976; Edward Rose to J. B. MacFarlane, 5 Aug. 1976; Edward Rose to David Mathews, 12 July 1976; all box 218, folder 3, Chancellor's papers, UCB.

104. Donald Chalkley to Edward Rose, 28 Sept. 1976, box 219, folder 3, Chancellor's papers, UCB; Chalkley to Berry, 25 Aug. 1976, UCB.

105. Omar Bartos and Robert Hanson to Committee for Reconciliation of Academic Freedom and Public Law 93-348, 23 Nov. 1976, box 219, folder 3, Chancellor's papers, UCB.

106. William Hodges to Charles Kenevan, 24 Mar. 1977, box 5, folder 10, NC-GTU.

107. Howard Higman to Milton Lipetz, 12 Apr. 1977, box 219, folder 3, Chancellor's papers, UCB.

108. Hodges to Kenevan, 24 Mar. 1977, NC-GTU.

109. Milton Lipetz to Carole Anderson, 4 Apr. 1977, box 219, folder 3, Chancellor's papers, UCB.

110. W. F. Hodges to members of the Faculty Council, 28 Apr. 1977, box 30, folder 5, Faculty Council papers, UCB.

111. "Nelson, Faculty Group to Meet on Research," *Silver and Gold* (27 June 1977), box 30, folder 5, Faculty Council papers, UCB.

112. *Crane v. Mathews*, 417 F. Supp. 532; 1976 U.S. Dist. LEXIS 16807.

113. DHEW, "Secretary's Interpretation of 'Subject at Risk,'" *Federal Register* 41 (28 June 1976): 26572.

114. Donald Chalkley to Philip G. Vargas, 6 Oct. 1976, box 219, folder 3, Chancellor's papers, UCB.

115. Chalkley to Rose, 28 Sept. 1976, UCB.

CHAPTER 3: THE NATIONAL COMMISSION

1. Robert J. Levine, Charles R. McCarthy, Edward L. Pattullo, and Kenneth J. Gergen, panelists, "The Political, Legal, and Moral Limits to Institutional Review Board (IRB) Oversight of Behavioral and Social Science Research," in Paula Knudson, ed., *PRIM&R through the Years: Three Decades of Protecting Human Subjects, 1974–2005* (Boston: Public Responsibility in Medicine & Research, 2006), 33.

2. Bradford H. Gray, "Human Subjects Review Committees and Social Research," in Murray L. Wax and Joan Cassell, *Federal Regulations: Ethical Issues and Social Research* (Boulder, CO: Westview, 1979), 47.

3. Dorothy I. Height, interview by Bernard A. Schwetz, 30 June 2004; Karen Lebacqz, interview by LeRoy B. Walters, 26 Oct. 2004; Tom Lamar Beauchamp, interview by Bernard A. Schwetz, 22 Sept. 2004; all in Oral History of the *Belmont Report*.

4. *National Research Act*, § 205.

5. *National Research Act*, § 202; Michael S. Yesley to Charles U. Lowe, 29 Nov. 1974, box 1, meeting #1, NCPHS-GU.

6. *National Research Act*, § 201.

7. The commission was based in part on a 1968 proposal by Senator Walter Mondale, who called upon "representatives of medicine, law, science, theology, philosophy, ethics, health administration, and government" to study "the ethical, legal, social, and political implications of biomedical advances." At that point, William Curran, professor of Health Law at Harvard University, had requested Mondale to "also include social scientists and theologians who have done significant work in the field." Thus the first proposal for including social scientists clearly saw them as more akin to theologians than to physicians. Senate Committee on Government Operations, *National Commission on Health Science and Society*, 90th Cong., 2nd sess., 1968, 1, 498.

8. *Community Mental Health Centers Act*, Public Law 95-622, § 1801.

9. Tom L. Beauchamp, interview by the author, Washington, DC, 18 Oct. 2007.

10. Bradford Gray, interview by the author, Washington, DC, 7 Aug. 2007.

11. Jonsen, interview by the author.

12. "ADAMHA's Suggestions for Organizations to Be Contacted in Regard to Nominations for the National Commission for the Protection of Human Subjects," Aug. 1974, Res 3-4, National Comm. for the Protection of Human Sub., folder #1, 1974–1975, RG 443.

13. DHEW, press release, 13 Sept. 1974, Res 3-4, National Comm. for the Protection of Human Sub., folder #1, 1974–1975, RG 443.

14. In June 1976, a researcher at the National Institute of Mental Health complained of this lack of a definition. See Arthur K. Leabman, Center for Studies of Crime and Delinquency, NIMH, to Natalie Reatig, 11 June 1976, box 8, meeting #19, tabs 9–11, NCPHS-GU. The following month, consultant Robert Levine suggested that behavioral research could be defined as "research having the purpose of understanding or modifying behavior," a definition expansive enough to cover the writing of novels. Robert J. Levine, "Similarities and Differences between Biomedical and Behavioral Research," 16 July 1976, box 9, meeting #21, tabs 6–8, NCPHS-GU.

15. National Commission, transcript, meeting #8, June 1975, 178, NCPHS-GU.

16. Don A. Gallant, "Response to Commission Duties as Detailed in PL 93-348," 4, box 5, meeting #15(A), tabs 10–11, NCPHS-GU.

17. "Prison Inmate Involvement in Biomedical and Behavioral Research in State Correctional Facilities," 31 Oct. 1975, box 3, meeting #12, tabs 1–8, NCPHS-GU.

18. *National Research Act*, § 202.

19. "Presentation to the Commission by R. Levine," draft, 27 Nov. 1974, box 1, meeting #1, NCPHS-GU.

20. National Commission, transcript, meeting #1, Dec. 1974, 2-118, NCPHS-GU.

21. National Commission, transcript, meeting #8, June 1975, 233–236, NCPHS-GU.

22. "The Boundaries between Biomedical or Behavioral Research and the Accepted and Routine Practice of Medicine," 14 July 1975, box 2, meeting #9, tabs 1–5, NCPHS-GU.

23. DHEW, "Protection of Human Subjects," 18917.

24. Robert J. Levine, "The Role of Assessment of Risk-Benefit Criteria in the Determination of the Appropriateness of Research Involving Human Subjects," preliminary draft, 2 Sept. 1975, box 3, meeting #10, NCPHS-GU.

25. Robert J. Levine, "Similarities and Differences between Biomedical and Behavioral Research," 16 July 1976, box 9, meeting #21, tabs 6–8, NCPHS-GU.

26. For very brief mentions, see National Commission, transcript, meeting #8, June 1975, 236, 242, NCPHS-GU.

27. National Commission, transcript, meeting #15, Feb. 1976, 317–324, NCPHS-GU.

28. Bradford H. Gray, *Human Subjects in Medical Experimentation* (New York: John Wiley & Sons, 1975), 237–250.

29. Gray, interview by the author.

30. Gray, interview by the author.

31. Paul Nejelski to Charles U. Lowe, 19 May 1975; Michael S. Yesley to Paul Nejelski, 26 Sept. 1975; both in box 3, meeting #11, tabs 8–11, NCPHS-GU. The book was eventually published as Paul Nejelski, ed., *Social Research in Conflict with Law and Ethics* (Cambridge, MA: Ballinger, 1976).

32. Bradford Gray to George Annas, Leonard Glantz, and Barbara Katz, 28 June 1976, box 9, meeting #21, tab 11, NCPHS-GU.

33. George Annas and Leonard Glantz to Bradford Gray, 7 July 1976, box 9, meeting #21, tab 11, NCPHS-GU.

34. Gray, interview by the author.

35. Barbara Mishkin, interview by the author, Chevy Chase, Maryland, 4 Oct. 2007.

36. Gray to Annas, Glantz, and Katz, 28 June 1976, NCPHS-GU.

37. National Commission, transcript, meeting #41, Apr. 1978, II-5, NCPHS-GU.

38. National Commission, transcript, meeting #40, Mar. 1978, 48, NCPHS-GU.

39. National Commission, transcript, meeting #38, Jan. 1978, 198, NCPHS-GU.

40. National Commission, transcript, meeting #41, Apr. 1978, II-10, NCPHS-GU.

41. Beauchamp, interview by the author.

42. *National Research Act*, § 202(1)(B)(v).

43. DHEW, *Institutional Guide*, 1; Phillips, "DHEW Regulations"; Bernard Barber, "Some Perspectives on the Role of Assessment of Risk/Benefit Criteria in the Determination of the Appropriateness of Research Involving Human Subjects," Dec. 1975, box 5, meeting #15(A), tabs 14–17, NCPHS-GU.

44. National Commission, transcript, meeting #8, June 1975, 241, NCPHS-GU.

45. Bradford H. Gray to Jerry A. Schneider, 16 Mar. 1976, box 7, meeting #17, tab 22, NCPHS-GU.

46. DHEW, "Institutional Review Boards: Report and Recommendations of the National Commission for the Protection of Human Subjects of Biomedical and Behavioral Research," *Federal Register* 43 (30 Nov. 1978): 56186.

47. Gray, interview by the author.

48. Survey Research Center, "Researcher Investigator Questionnaire," Winter 1976, 27, box 22, subject file: Instruments from IRB study, NCPHS 1976, tabs 1–3, NCPHS-GU.

49. Bradford H. Gray, Robert A. Cooke, and Arnold S. Tannenbaum, "Research Involving Human Subjects," *Science*, n.s., 201 (22 Sept. 1978): 1094–1101; Gray, interview by the author.

50. President's Commission, transcript of proceedings, 12 July 1980, 315, box 37, PCSEP-GU; Gray, interview by the author.

51. Survey Research Center, "Researcher Investigator Questionnaire," 5–6; Survey Research Center, "Research Involving Human Subjects," tables, 2 Oct. 1976, box 11, meeting #23, tab 3(b); both NCPHS-GU.

52. Survey Research Center, "Research Involving Human Subjects," table I.2, NCPHS-GU.

53. Robert A. Cooke to Bradford Gray, 3 Nov. 1976, table 18, box 11, meeting #24, tabs 7–9, NCPHS-GU.

54. Survey Research Center, "Self-Administered Form for Research Investigator," Winter 1976, 23, box 22, subject file: Instruments from IRB study, NCPHS 1976, tabs 1–3, NCPHS-GU.

55. Survey Research Center, "Research Involving Human Subjects," table II.2, NCPHS-GU.

56. Survey Research Center, "Research Involving Human Subjects," NCPHS-GU.

57. E. L. Pattullo, "Modesty Is the Best Policy: The Federal Role in Social Research," in Beauchamp et al., *Ethical Issues in Social Science Research*, 382. The same could be said for studies conducted by the federal government itself. In 1979, officials within DHEW found that most surveys were conducted without IRB review, "and there is no evidence of any adverse consequences." "Applicability to Social and Educational Research," attached to Peter Hamilton, Deputy General Counsel, to Mary Berry et al., 27 Mar. 1979, Res 3-1-B, Proposed Policy Protections Human Subjects 1978–79, RG 443.

58. Edward Rose to F. David Mathews, 12 July 1976, box 10, meeting #22, tab 12, NCPHS-GU.

59. Phyllis L. Fleming, Penelope L. Maza, and Harvey A. Moore to Bradford Gray, 3 Mar. 1976, box 7, meeting #17, tab 22, NCPHS-GU.

60. Roy. G. Francis to "Dear Chairperson," 23 Feb. 1976, box 7, meeting #17, tab 22, NCPHS-GU.

61. Murray Wax to Bradford Gray, 23 Apr. 1976, box 8, meeting #18, tab 14, NCPHS-GU.

62. Murray Wax to Bradford Gray, 18 June 1976, box 9, meeting #20, tabs 6–7, NCPHS-GU.

63. National Commission, transcript, meeting #23, Oct. 1976, 2-51, 2-76, NCPHS-GU.

64. "Mechanisms for Applying Ethical Principles to the Conduct of Research Involving Human Subjects: The Institutional Review Board," preliminary draft, 5 Nov. 1976, 36, box 11, meeting #24, tabs 5–6, NCPHS-GU.

65. Staff paper, "Social Research," 3 Dec. 1976, box 12, meeting #25, tabs 7–9, NCPHS-GU.

66. National Commission, transcript, meeting #25, Dec. 1976, 15–26, NCPHS-GU.

67. National Commission, *Appendix to Report and Recommendations: Institutional Review Boards*, DHEW Publication No. (OS) 78-0009 (Washington, DC: U.S. Department of Health, Education, and Welfare, 1978), testimony of John Clausen, Wallace Gingerich, Howard Higman, Ada Jacox, Hans Mauksch, Virginia Olesen, Edward Rose, William Sturtevant, and Linda Wilson.

68. National Commission, *Appendix to Report and Recommendations*, 3-15.

69. National Commission, transcript of public hearings on institutional review boards, 1977, 339–341, box 10, folder 4, NC-GTU.

70. National Commission, transcript of public hearings, 72, NC-GTU.

71. National Commission, transcript of public hearings, 243, NC-GTU.

72. National Commission, transcript of public hearings, 72, NC-GTU.

73. National Commission, transcript of public hearings, 135, NC-GTU.

74. National Commission, transcript of public hearings, 655–657, NC-GTU.

75. University of Wisconsin–Milwaukee, "Performance of the Institutional Review Board," 5 Apr. 1977, III-3, box 5, folder 14, NC-GTU.

76. Murray L. Wax, "On Fieldworkers and Those Exposed to Fieldwork: Federal Regulations, Moral Issues, Rights of Inquiry," Sept. 1976, box 11, meeting #23, tabs 7–8, NCPHS-GU.

77. Fleming, Maza, and Moore to Gray, 3 Mar. 1976; Rose to Mathews, 12 July 1976; both NCPHS-GU; Paul Kay to National Commission, 28 Feb. 1977, box 5, folder 10, NC-GTU.

78. National Commission, transcript of public hearings, 337, NC-GTU.

79. Herbert P. Phillips to Michael Yesley, 5 Aug. 1975, box 3, meeting #10, tabs 13–14; Phillips, "DHEW Regulations"; both NCPHS-GU.

80. Joe Shelby Cecil, "Summary of Reactions to the April, 1977 Third Draft of 'Protection of the Rights and Interests of Human Subjects in the Areas of Program Evaluation, Social Experimentation, and Statistical Analysis of Data from Administrative Records,'" Jan. 1978, box 20, meeting #40, tab 10, NCPHS-GU.

81. National Commission, transcript, meeting #31, June 1977, 106–109, NCPHS-GU.

82. "Institutional Review Boards," draft report, 1 July 1977, 33, 53, box 16, meeting #32, tab 2, NCPHS-GU.

83. "Issues Relating to the Performance of Institutional Review Boards," ca. July 1977, box 16, meeting #32, tab 2, NCPHS-GU.

84. "Institutional Review Boards," draft report, 42–43, NCPHS-GU.

85. National Commission, transcript, meeting #32, July 1977, 132, NCPHS-GU.

86. *Crane v. Mathews.*

87. National Commission, transcript, meeting #32, July 1977, 134–139, NCPHS-GU.

88. "Definition of a Human Subject," 2 Aug. 1977, box 17, meeting #33, tabs 8–9, NCPHS-GU.

89. "Institutional Review Boards," draft report, NCPHS-GU.

90. National Commission, transcript, meeting #33, Aug. 1977, 182, NCPHS-GU.

91. Gray, interview by the author.

92. National Commission, transcript, meeting #33, Aug. 1977, 165–182, NCPHS-GU.

93. Staff draft, "IRB Recommendations," 2 Sept. 1977, box 17, meeting #34, tabs 1–2, NCPHS-GU.

94. Since at least May 1976, Gray had argued against considering the policy consequences of research as part of the human subjects review process. Bradford H. Gray, "The Functions of Human Subjects Review Committees," *American Journal of Psychiatry* 134 (Aug. 1977): 909. In his 2007 interview with the author, Gray noted that he had supported this provision over the objections of Mishkin.

95. Staff draft, "IRB Recommendations," NCPHS-GU.

96. Constance Row to Bradford Gray, 21 Oct. 1976, box 11, meeting #23, tabs 7–8, NCPHS-GU.

97. Staff draft, "IRB Recommendations," NCPHS-GU.

98. National Commission, transcript, meeting #23, Oct. 1976, 2-52–2-57, NCPHS-GU; Survey Research Center, "Research Involving Human Subjects," NCPHS-GU.

99. Phillips, interview by the author.

100. National Commission, transcript, meeting #36, Nov. 1977, II-176, NCPHS-GU.

101. National Commission, transcript, meeting #34, Sept. 1977, 30–31, 46–50, NCPHS-GU.

102. Linda S. Wilson to National Commission, 7 Nov. 1977, box 18, meeting #36, tabs 9–10 (incomplete), NCPHS-GU; National Commission, transcript, meeting #36, Nov. 1977, II-182, NCPHS-GU.

103. National Commission, transcript, meeting #36, Nov. 1977, II-89, NCPHS-GU.

104. National Commission, transcript, meeting #37, Dec. 1977, 49, NCPHS-GU.

105. National Commission, transcript, meeting #37, Dec. 1977, 85–109, and meeting #38, Jan. 1978, 147–174; both NCPHS-GU.

106. National Commission, transcript, meeting #37, Dec. 1977, II-2, NCPHS-GU.

107. Mishkin, interview by the author.

108. National Commission, transcript, meeting #37, Dec. 1977, 101, NCPHS-GU.

109. President's Commission, transcript of proceedings, 324.

110. National Commission, transcript, meeting #39, Feb. 1978, 44, NCPHS-GU.

111. National Commission, transcript, meeting #41, Apr. 1978, II-3–II-34, II-43–II-46, NCPHS-GU.

112. National Commission, transcript, meeting #41, Apr. 1978, II-46, NCPHS-GU.

113. DHEW, "Institutional Review Boards," 56179.

114. DHEW, "Institutional Review Boards," 56196.

115. DHEW, "Institutional Review Boards," 56176.

116. DHEW, "Institutional Review Boards," 56190.

117. National Commission, transcript, meeting #5, Apr. 1975, 587, NCPHS-GU.

118. Gallant, "Response to Commission Duties," NCPHS-GU; Albert Reiss Jr., "Selected Issues in Informed Consent and Confidentiality with Special Reference to Behavioral/Social Science Research/Inquiry, February 1, 1976," in National Commission, *The Belmont Report: Ethical Principles and Guidelines for the Protection of Human Subjects of Research; Appendix, Vol. 2*, DHEW Publication No. (OS) 78-0014 (Washington, DC: GPO, 1978), 25-13; Wax to Gray, 18 June 1976, NCPHS-GU.

CHAPTER 4: THE *BELMONT REPORT*

1. OHRP, "Federalwide Assurance (FWA) for the Protection of Human Subjects, 6 January 2005," www.hhs.gov/ohrp/humansubjects/assurance/filasurt.htm [accessed 29 Apr. 2008].

2. In 2008, I requested copies of all U.S. university assurances that did *not* pledge to follow the *Belmont Report.* I received only two: from Northeast Iowa Community College and from Langston University. Carol Maloney to the author, 15 July 2008.

3. Mary Simmerling, Brian Schwegler, Joan E. Sieber, and James Lindgren, "Introducing a New Paradigm for Ethical Research in the Social, Behavioral, and Biomedical Sciences: Part I," *Northwestern University Law Review* 101 (Special Issue, 2007): 838–839.

4. Jonathan Kimmelman, "Review of *Belmont Revisited: Ethical Principles for Research with Human Subjects,*" *JAMA: Journal of the American Medical Association* 296 (2 Aug. 2006): 589.

5. The U.S. Constitution, of course, contains provisions for its own amendment. Other research ethics documents, such as the Declaration of Helsinki, do not, but their corporate authors survive to make revisions as needed. In contrast, the *Belmont Report* was written by a temporary commission which cannot be reconvened without an act of Congress.

6. James F. Childress, Eric M. Meslin, and Harold T. Shapiro, eds., *Belmont Revisited: Ethical Principles for Research with Human Subjects* (Washington, DC: Georgetown University Press, 2005).

7. Simmerling et al., "Introducing a New Paradigm," 840–841.

8. *National Research Act,* § 202(a)(A).

9. National Commission, *Report and Recommendations: Research on the Fetus,* DHEW Publication No. (OS) 76-127 (Washington, DC: U.S. Department of Health, Education, and Welfare, 1975), 63.

10. LeRoy Walters, "Some Ethical Issues in Research Involving Human Subjects," 2, box 4, meeting #15(A), tabs 7–9, NCPHS-GU.

11. H. Tristam Engelhardt Jr., "Basic Ethical Principles in the Conduct of Biomedical and Behavioral Research Involving Human Subjects," Dec. 1975, 9, box 4, meeting #15(A), tabs 5–6, NCPHS-GU.

12. "Ethical Principles and Human Experimentation," 19 Jan. 1976, 13, box 4, meeting #15(A), tabs 1–2, NCPHS-GU.

13. Richard A. Tropp, "What Problems Are Raised When the Current DHEW Regulation on Protection of Human Subjects Is Applied to Social Science Research?" 13 Feb. 1976, 16, box 5, meeting #15(A), Miscellaneous Memoranda & Reports, NCPHS-GU. Reprinted in National Commission, *Belmont Report: Appendix, Vol. 2.*

14. Bernard Barber, John J. Lally, Julia Loughlin Makarushka, and Daniel Sullivan, *Research on Human Subjects: Problems of Social Control in Medical Experimentation,* rev. ed. (New Brunswick, NJ: Transaction, 1979), xvi n. 2.

15. Barber, "Some Perspectives on the Role of Assessment of Risk/Benefit Criteria."

16. National Commission, *Appendix to Report and Recommendations: Institutional Review Boards,* 3-36–3-37.

17. Donald J. Black and Albert J. Reiss Jr., "Police Control of Juveniles," *American Sociological Review* 35 (Feb. 1970): 65.

18. Albert J. Reiss Jr., "Selected Issues in Informed Consent and Confidentiality, with Special Reference to Behavioral/Social Science Research/Inquiry," 1 Feb. 1976, 34–36, 42, 58, 161, box 5, meeting #15(A), tab 23, NCPHS-GU. Reiss borrowed the term "muckraking sociology" from an anthology of that title edited by Gary Marx.

19. Shils, "Muting the Social Sciences at Berkeley," 290–295.

20. National Commission, transcript, meeting #15, Feb. 1976, 14, 118, NCPHS-GU.

21. National Commission, transcript, meeting #15, Feb. 1976, 260, NCPHS-GU.

22. National Commission, transcript, meeting #15, Feb. 1976, 232, NCPHS-GU.

23. National Commission, transcript, meeting #15, Feb. 1976, 73, NCPHS-GU.

24. National Commission, transcript, meeting #15, Feb. 1976, 14, NCPHS-GU.

25. National Commission, transcript, meeting #15, Feb. 1976, 313, NCPHS-GU.

26. "Identification of Basic Ethical Principles," 1 Mar. 1976, box 6, meeting #16, tabs 4–5, NCPHS-GU; for Toulmin's authorship, see National Commission, transcript, meeting #15, Feb. 1976, 316, NCPHS-GU.

27. National Commission, transcript, meeting #15, Feb. 1976, 315, NCPHS-GU.

28. "Identification of Basic Ethical Principles," NCPHS-GU.

29. Reiss, "Selected Issues in Informed Consent and Confidentiality," 25–66, NCPHS-GU.

30. Leslie Kish, *Survey Sampling* (New York: John Wiley & Sons, 1965), 408.

31. Beauchamp, interview by the author.

32. Tom L. Beauchamp, "The Origins and Evolution of the *Belmont Report*," in Childress et al., *Belmont Revisited*, 15.

33. Beauchamp, interview by the author.

34. Beauchamp, interview by the author.

35. American Psychological Association, *Ethical Principles*.

36. American Psychological Association, *Ethical Principles*, 1–2.

37. National Commission, transcript, meeting #41, Apr. 1978, II-25; Beauchamp, interview by the author.

38. Beauchamp, interview by the author.

39. National Commission, transcript, meeting #37, Dec. 1977, 104, NCPHS-GU. In 1986, Beauchamp coauthored a book that acknowledged "methodological differences" and "different perspectives" on ethics among the various social and behavioral sciences, but he still privileged psychology as having the "lengthiest history of struggle with the problem of consent within the behavioral and social sciences." Ruth R. Faden and Tom L. Beauchamp, *A History and Theory of Informed Consent* (New York: Oxford University Press, 1986), 167.

40. Beauchamp, interview by the author.

41. "Discussion Paper: Some Issues Not Fully Resolved at Belmont," 4 Feb. 1977, box 12, meeting #27, tabs 2–3, NCPHS-GU.

42. "Belmont Paper: Ethical Principles for Research Involving Human Subjects," draft, 1 Apr. 1977, box 15, meeting #30, tabs 1–2, NCPHS-GU.

43. Beauchamp later wrote that "for behavioral research, I started with Stuart E. Golann, 'Ethical Standards for Psychology: Development and Revisions, 1938–1968,' *Annals*

of the New York Academy of Sciences 169 (1970): 398–405, and American Psychological Association, Inc., *Ethical Principles in the Conduct of Research with Human Participants* (Washington, DC: APA, 1973)." There is no indication he read anything by social scientists outside of psychology. Beauchamp, "The Origins and Evolution of the *Belmont Report,*" 22 n. 10.

44. Nicholas Hobbs, "The Development of a Code of Ethical Standards for Psychology," *American Psychologist* 3 (Mar. 1948): 80–84; American Psychological Association, *Ethical Principles*, 3.

45. "Belmont Paper," draft, 1 Apr. 1977, NCPHS-GU.

46. Jonsen, *Birth of Bioethics*, 103; Jonsen, "On the Origins and Future of the *Belmont Report,*" in Childress et al., *Belmont Revisited*, 5.

47. "Belmont Paper: Ethical Principles for Research Involving Human Subjects," draft, 2 Dec. 1977, box 19, meeting #38, tabs 2–3, NCPHS-GU.

48. Beauchamp, interview by the author.

49. Jonsen, interview by the author.

50. *National Research Act*, § 202 (B).

51. Beauchamp, interview by the author.

52. "Belmont Paper," draft, 2 Dec. 1977, 15, NCPHS-GU.

53. Beauchamp, interview by the author.

54. "Belmont Paper," draft, 2 Dec. 1977, 20, NCPHS-GU.

55. Reiss, "Selected Issues in Informed Consent and Confidentiality," 25–42, NCPHS-GU.

56. "Belmont Paper," draft, 2 Dec. 1977, 28, NCPHS-GU.

57. The Society for Applied Anthropology did recommend that members not advocate *action* that was not beneficial, but it did not require such a test for *research*. Society for Applied Anthropology, "Statement on Professional and Ethical Responsibilities," 13 Mar. 1974, box 8, meeting #18, tab 14, NCPHS-GU.

58. Carl B. Klockars, "Field Ethics for the Life History," in Robert Weppner, ed., *Street Ethnography: Selected Studies of Crime and Drug Use in Natural Settings* (Beverly Hills, CA: Sage, 1977), 222.

59. Reiss, "Selected Issues in Informed Consent and Confidentiality," 25–69, NCPHS-GU.

60. Jonsen, "On the Origins and Future of the *Belmont Report,*" in Childress et al., *Belmont Revisited*, 8.

61. National Commission, *The Belmont Report: Ethical Principles and Guidelines for the Protection of Human Subjects of Research*, DHEW Publication No. (OS) 78-0012 (Washington, DC: GPO, 1978).

62. National Commission, transcript, meeting #39, Feb. 1978, II-40, NCPHS-GU.

63. National Commission, transcript, meeting #39, Feb. 1978, II-92, NCPHS-GU.

64. National Commission, transcript, meeting #40, Mar. 1978, II-13, NCPHS-GU.

65. National Commission, transcript, meeting #40, Mar. 1978, II-35–II-56, NCPHS-GU.

66. National Commission, transcript, meeting #39, Feb. 1978, II-61–II-63, NCPHS-GU.

67. National Commission, transcript, meeting #40, Mar. 1978, II-12–II-14, NCPHS-GU.

68. "Belmont Paper: Ethical Principles and Guidelines for Research Involving Human Subjects," draft, 6 Apr. 1978, 4, box 20, meeting #41, tabs 2–3, NCPHS-GU.

69. "Belmont Paper: Ethical Principles for Research Involving Human Subjects," 6 July 1978, box 3, folder 3, NC-GTU; "The Belmont Report: Ethical Guidelines for the Protection of Human Subjects of Research," draft, 8 Sept. 1978, box 21, meeting #43, tabs 1–2, NCPHS-GU.

70. Levine et al., "Political, Legal, and Moral Limits," in Knudson, *PRIM&R through the Years*, 33.

71. Finbarr W. O'Connor, "The Ethical Demands of the *Belmont Report*," in Carl B. Klockars and Finbarr W. O'Connor, eds., *Deviance and Decency: The Ethics of Research with Human Subjects* (Beverly Hills, CA: Sage, 1979), 241.

72. National Commission, *Belmont Report*, 3.

73. "IRB Recommendations," draft, 12 Jan. 1978, 10, box 19, meeting #38, tabs 2–3, NCPHS-GU.

74. National Commission, transcript, meeting #38, Jan. 1978, 162, NCPHS-GU.

75. "Belmont Paper: Ethical Principles and Guidelines for Research Involving Human Subjects," draft, 3 Feb. 1978, 2, box 20, meeting #39, tabs 3–4, NCPHS-GU.

76. "IRB Recommendations," draft, 12 Jan. 1978, 10–13, NCPHS-GU.

77. DHEW, "Institutional Review Boards," 56179–56180.

78. The appendices should not be taken as a record of what the commissioners had on hand at Belmont. Donald T. Campbell and Joe Shelby Cecil's paper on program evaluation, in particular, was "completely rewritten" between January 1976 and January 1977, and it is the latter version that was printed in the appendix. See Yesley to Commission Members, 10 Feb. 1977, box 13, meeting #27, tabs 4–7, NCPHS-GU.

79. Albert J. Reiss Jr., "Governmental Regulation of Scientific Inquiry: Some Paradoxical Consequences," in Klockars and O'Connor, *Deviance and Decency*.

80. Jonsen, interview by the author.

81. Beauchamp, interview by the author.

82. "Identification of Basic Ethical Principles," NCPHS-GU.

83. Beauchamp, interview by the author.

84. Frederick G. Hofmann to Kenneth J. Ryan, 22 Jan. 1976, box 5, folder 26, NC-GTU.

CHAPTER 5: THE BATTLE FOR SOCIAL SCIENCE

1. Arthur Herschman, "Science and Technology: New Tools, New Dimensions; AAAS Annual Meeting, 12–17 February 1978, Washington," *Science* 198 (4 Nov. 1977): 493; Wax and Cassell, *Federal Regulations*.

2. Bradford H. Gray, "Comment," *American Sociologist* 13 (Aug. 1978): 161.

3. Thomas A. Bartlett, Robert Clodius, and Sheldon Steinbach to Charles McCarthy, 29 Jan. 1979, box 52, MIT Committee 2/2, IdSP.

4. Daniel Patrick Moynihan to Joseph Califano, 17 Jan. 1979, box 25, Human Subjects Recent Information Part II, IdSP.

5. McCarthy, interview by El-Hinnawy.

6. Charles McCarthy and Donna Spiegler to Joel M. Mangel and Richard A. Tropp, 2 Aug. 1978, Res 3-1. Human Subjects Policy & Regulations 1973–82, RG 443.

7. Hamilton to Berry et al., 27 Mar. 1979, RG 443.

8. DHEW, "Institutional Review Boards," 56175.

9. Hamilton to Berry et al., 27 Mar. 1979, RG 443.

10. Dean Gallant to William Dommel, 5 Oct. 1979, box 24, Human Subjects Corresp. 1979 1/3, IdSP.

11. DHEW, draft memo to the Secretary, ca. Sept. 1978, box 26, Human Subjects Basic Documents, IdSP.

12. Richard Tropp to Ithiel de Sola Pool, 4 Oct. 1979, box 26, Human Subjects Basic Documents, IdSP.

13. DHEW, draft memo to the Secretary, ca. Sept. 1978, IdSP.

14. "Applicability to Social and Educational Research," RG 443.

15. "Applicability to Social and Educational Research," RG 443.

16. Gerald L. Klerman, Administrator, to Assistant Secretary for Health and Surgeon General, 30 Mar. 1979, Res 3-1-B Proposed Policy Protections Human Subjects 1978–79, RG 443.

17. Julius B. Richmond to Acting General Counsel, draft, 12 June 1979, Res 3-1-B Proposed Policy Protections Human Subjects 1978–79, RG 443.

18. Richmond to Acting General Counsel, draft, 12 June 1979, RG 443.

19. "Applicability to Social and Educational Research," RG 443.

20. Richmond to Acting General Counsel, draft, 12 June 1979, RG 443.

21. Hamilton to Mary Berry et al., 27 Mar. 1979, RG 443.

22. Donald F. Frederickson, Director, NIH, to Assistant Secretary for Health and Surgeon General, 18 Apr. 1979, Res 3-1-B Proposed Policy Protections Human Subjects 1978–79, RG 443.

23. Julius Richmond, Assistant Secretary for Health and Surgeon General to Deputy General Counsel, draft memo, attached to Frederickson to Assistant Secretary for Health and Surgeon General, 18 Apr. 1979, RG 443.

24. Gerald L. Klerman to Dick Beattie, Rick Cotton, and Peter Hamilton, 11 June 1979, Res 3-1-B Proposed Policy Protections Human Subjects 1978–79, RG 443.

25. Klerman to the Assistant Secretary, 30 Mar. 1979; Peter B. Hamilton to the Secretary, draft memo, 4 June 1979, Res 3-1-B Proposed Policy Protections Human Subjects 1978–79; both in RG 443.

26. Charles R. McCarthy, comments on the panel "The Political, Legal, and Moral Limits to Institutional Review Board (IRB) Oversight of Behavioral and Social Science Research," in Knudson, *PRIM&R through the Years*, 36.

27. DHEW, "Proposed Regulations Amending Basic HEW Policy for Protection of Human Research Subjects," *Federal Register* 44 (14 Aug. 1979): 47693.

28. DHEW, "Proposed Regulations Amending Basic HEW Policy," 47688.

29. Ithiel de Sola Pool, "The Necessity for Social Scientists Doing Research for Governments," in Horowitz, *Rise and Fall of Project Camelot*, 268.

30. Pool, "Necessity for Social Scientists Doing Research for Governments," 267.

31. Ithiel de Sola Pool to *Harvard Crimson*, 6 Oct. 1969, box 180, Letter to *Harvard Crimson*, 1969, IdSP.

32. Ithiel de Sola Pool to F. William Dommel, 8 Nov. 1979, box 24, Human Subjects Mailings 2/4, IdSP; Ithiel de Sola Pool, "Censoring Research," *Society* (Nov./Dec. 1980): 40.

33. Ithiel Pool [*sic*], "Draft Statement on Procedures and Guidelines," May 1975, box 52, MIT Committee 1/2, IdSP.

34. Ithiel de Sola Pool to Myron [Weiner?], 20 Oct. 1975, box 52, MIT Committee 1/2, IdSP.

35. Ithiel de Sola Pool, "Human Subjects Regulations on the Social Sciences," presented at the New York Academy of Sciences, 6 May 1981, box 24, Human Subjects Mailings 1/4, IdSP.

36. Pool to Myron, IdSP.

37. Ithiel de Sola Pool to "To Whom It May Concern," 8 Dec. 1978, box 26, Human Subjects Thank You Letters 1/2, IdSP.

38. Ithiel de Sola Pool to John S. Nichols, 29 Dec. 1979, box 25, Human Subjects Corresp. 1979 2/3, IdSP.

39. Pool, "Censoring Research," 40.

40. Ithiel de Sola Pool to Philip Abelson, 30 Dec. 1979, box 25, Human Subjects Corresp. 1979, IdSP.

41. Anthony Lewis to Ithiel de Sola Pool, 5 Jan. 1979, box 26, Human Subjects Letters 1/2, IdSP.

42. Anthony Lewis, "Pettifog on the Potomac," *New York Times*, 15 Jan. 1979.

43. Correspondence in box 24, Human Subjects Mailings 2/4, IdSP.

44. Antonin Scalia to Ithiel de Sola Pool, 24 Jan. 1980, box 25, Human Subjects Corresp. 1980 3/3, IdSP.

45. Ithiel de Sola Pool to Irving Horowitz, 28 Nov. 1980, box 25, Human Subjects Corresp. 1980 2/3, IdSP.

46. Ithiel de Sola Pool to "Committee of Concern," 9 Sept. 1979, box 24, Human Subjects Mailings 4/4, IdSP.

47. J. W. Peltason to "Parties Concerned with Federal IRB Regulations," 3 Oct. 1979, box 24, Human Subjects Mailings 2/4, IdSP.

48. University of California, Davis Department of Sociology to F. William Dommel Jr., 17 Oct. 1979, box 25, Human Subjects Corresp. 1979 2/3, IdSP.

49. Lauren H. Seiler to F. William Dommel Jr., 7 Nov. 1979, box 25, Human Subjects Corresp. 1979 3/3, IdSP. The paper was later published as Lauren H. Seiler and James M. Murtha, "Federal Regulation of Social Research," *Freedom at Issue* (Nov.–Dec. 1979): 26–30.

50. "Issues Related to HHS Human Subject Research," ca. 20 May 1980, Res 3-1-B Proposed Policy Protections Human Subjects 1979–80, RG 443.

51. Linda S. Wilson to F. William Dommel Jr., 5 Nov. 1979, box 24, Human Subjects Corresp. 1979 1/3, IdSP.

52. Seiler to Dommel, 7 Nov. 1979, IdSP.

53. E. L. Pattullo, e-mail to the author, 14 Aug. 2007.

54. E. L. Pattullo to Richard Tropp, 23 Aug. 1978, box 26, Human Subjects Letters 1/2, IdSP.

55. Fred Hiatt, "Watchdogs and Guinea Pigs," *Harvard Crimson*, 15 Dec. 1975.

56. E. L. Pattullo, "Comment," *American Sociologist* 13 (Aug. 1978): 168.

57. Pattullo to Tropp, 23 Aug. 1978, IdSP.

58. Ithiel de Sola Pool to E. L. Pattullo, 19 Sept. 1979, box 25, Human Subjects Corresp. 1979 2/3, IdSP.

59. Richard T. Louttit and Edward L. Pattullo, workshop facilitators, "When Is a Subject a Subject?" in Knudson, *PRIM&R through the Years*, 30.

60. Levine et al., "Political, Legal, and Moral Limits," in Knudson, *PRIM&R through the Years*, 38–40.

61. Levine et al., "Political, Legal, and Moral Limits," in Knudson, *PRIM&R through the Years*, 40.

62. John Ball to Gil Omenn, 21 Nov. 1979, box 24, Human Subjects Corresp. 1979 1/3, IdSP.

63. American Association of University Professors, "Regulations Governing Research on Human Subjects," *Academe* (Dec. 1981): 363. The full text is printed as J. W. Peltason, "Comment on the Proposed Regulations from Higher Education and Professional Social Science Associations," *IRB: Ethics and Human Research* 2 (Feb. 1980): 10.

64. Levine et al., "Political, Legal, and Moral Limits," in Knudson, *PRIM&R through the Years*, 33.

65. Director, OPRR, to Assistant Secretary for Health and Surgeon General, 15 Oct. 1979, Res 3-1-B Proposed Policy Protections Human Subjects 1979–80, RG 443.

66. Robert Levine, interview by Bernard A. Schwetz, 14 May 2004, Oral History of the *Belmont Report*; McCarthy, interview by El-Hinnawy.

67. Ball to Omenn, 21 Nov. 1979, IdSP.

68. Reiss, "Governmental Regulation of Scientific Inquiry," 67.

69. Beauchamp, interview by the author; Constance Holden, "Ethics in Social Science Research," *Science* 206 (2 Nov. 1979): 537–538, 540. The conference proceedings were eventually published, but only after the new regulations went into effect and therefore too late to shape policy. Beauchamp et al., *Ethical Issues in Social Science Research*.

70. Pat [E. L. Pattullo] to Ithiel [de Sola Pool], 18 June 1980, box 25, Human Subjects Corresp. 1980 2/3, IdSP.

71. Ithiel de Sola Pool to Louis Menand, Myron Weiner, and Alan Altshuler, 12 Nov. 1979, box 25, Human Subjects Corresp. 1979 2/3, IdSP.

72. Ithiel de Sola Pool, "Prior Restraint," *New York Times*, 16 Dec. 1979.

73. "H.E.W. Line," *Nation* (31 May 1980): 645; "The Guinea Pigs," *Wall Street Journal*, 27 May 1980.

74. Edward J. Markey to Ithiel de Sola Pool, 31 Oct. 1979; J. Kenneth Robinson to William R. Nelson, 3 Oct. 1979; Larry Pressler to Ithiel de Sola Pool, 4 Oct. 1979; all in box 24, Human Subjects Mailings 1/4; Paul Tsongas to Ithiel de Sola Pool, 6 Oct. 1979, box 25,

Human Subjects Corresp. 1979 3/3; Daniel Patrick Moynihan to Ithiel de Sola Pool, 11 June 1980, box 25, Human Subjects Corresp. 1980 2/3; all in IdSP.

75. House Committee on Interstate and Foreign Commerce, *Health Research Act of 1980: Report, Together with Additional and Minority Views, to Accompany H.R. 7036*, 96th Cong., 2nd sess., H. Report 96-997, 1980, 140.

76. House Committee on Interstate and Foreign Commerce, *Health Research Act of 1980*, 37, 138.

77. Ithiel de Sola Pool to "Committee of Concern," 5 June 1980, box 88, Material on Human Subjects, IdSP.

78. President's Commission, briefing book for public meeting, 12 July 1980, PCSEP-GU. The briefing book includes an essay by Pool and other materials that resulted from his campaign, including the *Nation* and *Wall Street Journal* editorials.

79. President's Commission, transcript of proceedings, 12 July 1980, PCSEP-GU.

80. President's Commission, transcript of proceedings, 12 July 1980, 333–340, PCSEP-GU.

81. President's Commission, transcript of proceedings, 12 July 1980, 273, 298, PCSEP-GU.

82. Morris Abram to Patricia Roberts Harris, Secretary, HHS, 18 Sept. 1980, Res 3-4, President's Commission for Study of Ethical Problems in Medicine and Research, folder #4 1978–80, RG 443.

83. Abram to Harris, 18 Sept. 1980, RG 443.

84. Abram to Harris, 18 Sept. 1980, RG 443.

85. James T. McIntyre Jr. and Frank Press to Patricia Roberts Harris, 26 Nov. 1980, Res 3-4 President's Commission for Study of Ethical Problems in Medicine and Research 1981, folder #1, RG 443.

86. Pat [E. L. Pattullo] to Ithiel [de Sola Pool], 31 Mar. 1980, box 25, Human Subjects Corresp. 1980 2/3, IdSP.

87. Parker Coddington to Pat Pattullo and Ithiel de Sola Pool, 17 Sept. 1980, box 24, Human Subjects Articles 1980, IdSP.

88. McCarthy, interview by El-Hinnawy.

89. Frederickson to the Secretary, 21 Nov. 1980, Res 3-1-B Proposed Policy Protections Human Subjects 1979–80, RG 443.

90. Charles R. McCarthy to Director, NIH, 20 Nov. 1980, Res 3-1-B Proposed Policy Protections Human Subjects 1979–80, RG 443.

91. Julius B. Richmond to the Secretary, draft, 25 Nov. 1980, Res 3-4 President's Commission for Study of Ethical Problems in Medicine and Research 1981, folder #1, RG 443.

92. "Applicability to Social and Educational Research," RG 443; Tropp, "What Problems Are Raised," in National Commission, *Belmont Report: Appendix, Vol. 2*, 18-1.

93. Joan Z. Bernstein, "The Human Research Subjects H.E.W. Wants to Protect," *New York Times*, 24 Jan. 1980; Pool, "Censoring Research," 39.

94. Chalkley to Berry, 25 Aug. 1976, UCB.

95. McCarthy to the Director, NIH, 20 Nov. 1980, RG 443.

96. Frederickson to the Secretary, 21 Nov. 1980, RG 443.

97. Alexander Capron, "IRBs: The Good News and the Bad News," in Knudson, *PRIM&R through the Years*, 72.

98. DHEW, "Institutional Review Boards," 56188.

99. Julius B. Richmond to the Secretary, 8 Jan. 1981, Res 3-4 President's Commission for Study of Ethical Problems in Medicine and Research 1981, folder #1, RG 443. The memo indicates that it was drafted by McCarthy on 29 Dec. 1980 and revised on Dec. 30.

100. McCarthy, interview by El-Hinnawy.

101. Charles R. McCarthy, "Introduction: The IRB and Social and Behavioral Research," in Joan E. Sieber, ed., *NIH Readings on the Protection of Human Subjects in Behavioral and Social Science Research: Conference Proceedings and Background Papers* (Frederick, MD: University Publications of America, 1984), 8–9.

102. DHEW, "Proposed Regulations Amending Basic HEW Policy," 47695.

103. Ithiel de Sola Pool, "Human Subjects Regulations on the Social Sciences," IdSP.

104. HHS, "Final Regulations Amending Basic HHS Policy for the Protection of Human Research Subjects," *Federal Register* 46 (26 Jan. 1981): 8369–8374.

105. Richard A. Tropp, "A Regulatory Perspective on Social Science Research," in Beauchamp et al., *Ethical Issues in Social Science Research*, 398.

106. Richard T. Louttit, "Government Regulations: Do They Facilitate or Hinder Social and Behavioral Research?" in Sieber, *NIH Readings*, 180.

107. Bettina Huber, "New Human Subjects Policies Announced: Exemptions Outlined," *American Sociological Association Footnotes* (Nov. 1981): 1.

108. Robert Reinhold, "New Rules for Human Research Appear to Answer Critics' Fear," *New York Times*, 22 Jan. 1981.

109. E. L. Pattullo, "How General an Assurance?" *IRB: Ethics and Human Research* 3 (May 1981): 8.

110. E. L. Pattullo, e-mail to the author, 14 Aug. 2007; E. L. Pattullo to Charles McCarthy, 30 Jan. 1981, Res 3-1, Human Subjects Policy & Regulations 1973–82, RG 443.

111. Ithiel de Sola Pool to members of the Committee of Concern about Human Subjects Regulations and other interested parties, 30 Jan. 1981, Res 3-1, Human Subjects Policy & Regulations 1973–82, RG 443.

112. Ithiel de Sola Pool to Arthur L. Caplan, 30 Dec. 1979, box 25, Human Subjects Corresp. 1979 1/3, IdSP.

CHAPTER 6: DÉTENTE AND CRACKDOWN

1. Senate Committee on Labor and Human Resources, *National Advisory Commission on Health Science and Society*, 59.

2. HHS, "Final Regulations Amending Basic HHS Policy," 8374.

3. OPRR, "Sample Multiple Project Assurance of Compliance with the Department of Health and Human Services' Regulation for the Protection of Human Subjects," 45 CFR 46, 3 July 1981, rev. 11 Aug. 1981, box 24, Human Subjects Mailings 1/4, IdSP.

4. Richmond to the Secretary, 8 Jan. 1981, RG 443.

5. Pattullo, "How General an Assurance?" 8–9.

6. E. L. Pattullo to Harvard University Committee on the Use of Human Subjects members, 22 July 1981, box 24, Human Subjects: (Materials Removed from Files for New Haven Talk), IdSP.

7. E. L. Pattullo to Charles McCarthy, 30 July 1981, box 24, Human Subjects: (Materials Removed from Files for New Haven Talk), IdSP.

8. Pool, "Human Subjects Regulations on the Social Sciences," IdSP.

9. Ithiel de Sola Pool, "To Sign or Not to Sign OPRR's General Assurances," *IRB: Ethics and Human Research* 3 (Dec. 1981): 8.

10. Charles R. McCarthy, reply to Ithiel de Sola Pool, "To Sign or Not to Sign OPRR's General Assurances," *IRB: Ethics and Human Research* 3 (Dec. 1981): 8–9.

11. E. L. Pattullo, "Governmental Regulation of the Investigation of Human Subjects in Social Research," *Minerva* 23 (Dec. 1985): 529.

12. Charles R. McCarthy to Assistant Secretary for Health and Surgeon General, 7 May 1980, Res 3-1-B Proposed Policy Protections Human Subjects 1979–80, RG 443.

13. McCarthy, interview by El-Hinnawy.

14. Lamont-Havers to Stone, 13 May 1974, RG 443.

15. Assistant Secretary for Health to Secretary, HHS, draft cover memo, Jan. 1984, RES 3-1-D Proposed Mondel [*sic*] Federal Policy Protection of Human Subjects, OD-NIH.

16. Joan P. Porter, interview by the author, Washington, DC, 2 Aug. 2007. In 2007 my Freedom of Information Act request for records of this phase was met with the reply that "no responsive records" exist.

17. "Responses of Department and Agencies, Model Policy and Other Recommendations," 16 Sept. 1983, O&M 2-N-4, Office for Protection from Research Risks (OPRR), OD-NIH.

18. "Concurrences of Departments and Agencies Including Proposed Departures from Model Policy," 3 May 1985, RES 3-1-D Proposed Mondel [*sic*] Federal Policy Protection of Human Subjects, OD-NIH.

19. "Concurrences of Departments and Agencies," 3 May 1985, OD-NIH.

20. Porter, interview by the author.

21. Model Federal Policy for Protection of Human Research Subjects, Jan. 1984, RES 3-1-D Proposed Mondel [*sic*] Federal Policy Protection of Human Subjects, OD-NIH.

22. Charles R. McCarthy to Gordon Wallace, 10 Apr. 1984, RES 3-1-D Proposed Mondel [*sic*] Federal Policy Protection of Human Subjects, OD-NIH.

23. Office of Science and Technology Policy, "Proposed Model Federal Policy for Protection of Human Subjects," *Federal Register* 51 (3 June 1986): 20206.

24. Porter, interview by the author.

25. Porter, interview by the author; "Federal Policy for the Protection of Human Subjects: Notice and Proposed Rules," *Federal Register* 53 (10 Nov. 1988): 45663, 45672.

26. Office of Science and Technology Policy et al., "Federal Policy for the Protection of Human Subjects: Notices and Rules," *Federal Register* 56 (18 June 1991): 28007.

27. Gary Ellis, interview by the author, Washington, DC, 19 Dec. 2008.

28. Citro et al., *Protecting Participants*, Appendix D.

29. "Communication Scholars' Narratives of IRB Experiences," *Journal of Applied Communication Research* 33 (Aug. 2005): 229.

30. H. Russell Bernard, *Research Methods in Cultural Anthropology: Qualitative and Quantitative Approaches* (Newbury Park, CA: Sage, 1988), 116.

31. H. Russell Bernard, *Research Methods in Cultural Anthropology: Qualitative and Quantitative Approaches*, 2nd ed. (Thousand Oaks, CA: Sage, 1994), 109.

32. H. Russell Bernard, *Research Methods in Anthropology: Qualitative and Quantitative Approaches*, 3d. ed. (Walnut Creek, CA: AltaMira, 2003), 138.

33. Fluehr-Lobban, *Ethics and the Profession of Anthropology*.

34. Carolyn Fluehr-Lobban, "Ethics and Professionalism in Anthropology: Tensions between Its Academic and Applied Branches," *Business and Professional Ethics* 10 (issue 4, 1991): 65.

35. Michael Dean Murphy and Agneta Johannsen, "Ethical Obligations and Federal Regulations in Ethnographic Research and Anthropological Education," *Human Organization* 49 (Summer 1990): 129–130.

36. Rik Scarce, *Contempt of Court: A Scholar's Battle for Free Speech from Behind Bars* (Lanham, CA: AltaMira, 2005), 13.

37. Alan Sica, "Sociology as a Worldview," *American Journal of Sociology* 102 (July 1996): 254; Christopher Shea, "Don't Talk to the Humans: The Crackdown on Social Science Research," *Lingua Franca* (Sept. 2000): 31.

38. Sudhir Venkatesh, *Gang Leader for a Day: A Rogue Sociologist Takes to the Streets* (New York: Penguin, 2008), 119, 203. Venkatesh believed that "the work of graduate students is largely overlooked." More likely, his experience reflects the time of his work (the early 1990s) rather than his student status.

39. Stefan Timmermans, "Cui Bono? Institutional Review Board Ethics and Ethnographic Research," *Studies in Symbolic Interaction* 19 (1995): 154.

40. Patricia A. Marshall, "Research Ethics in Applied Anthropology," *IRB: Ethics and Human Research* 14 (Nov.–Dec. 1992): 5 n. 37.

41. American Sociological Association, "Index to Issues and Articles, *Footnotes*," www.asanet.org [accessed 23 June 2008].

42. Robert E. Cleary, "The Impact of IRBs on Political Science Research," *IRB: Ethics and Human Research* 9 (May–June 1987): 6–10.

43. Committee on Professional Ethics, Rights, and Freedoms, "Resolution on Ethics and Human Subjects," *PS: Political Science and Politics* 22 (Mar. 1989): 106.

44. American Sociological Association, *Code of Ethics* (Washington, DC: American Sociological Association, 1989), 3.

45. American Association of University Professors, "Regulations Governing Research on Human Subjects," 368.

46. Keith Schneider, "Secret Nuclear Research on People Comes to Light," *New York Times*, 17 Dec. 1993; editorial, "Experiments on Humans," *Washington Post*, 20 Dec. 1993.

47. John H. Cushman, "Study Sought on All Testing on Humans," *New York Times*, 10 Jan. 1994.

48. Department of Defense, "Report on Search for Human Radiation Experiment Records 1944–1994," appendix 2, exhibit 4, www.defenselink.mil/pubs/dodhre/ [accessed 22 Dec. 2008].

49. ACHRE, *Final Report* (New York: Oxford University Press, 1996), 523–524.

50. NBAC, *1996–1997 Annual Report* (Rockville, MD: National Bioethics Advisory Commission, 1998), 3.

51. General Accounting Office, *Scientific Research: Continued Vigilance Critical to Protecting Human Subjects* (Washington, DC: General Accounting Office, 1996), 24.

52. NBAC, *Reports and Recommendations of the National Bioethics Advisory Commission*, vol. 1 of *Ethical and Policy Issues in Research Involving Human Participants* (Washington, DC: National Bioethics Advisory Commission, 2001), 143.

53. General Accounting Office, *Scientific Research*, 2.

54. HHS, *Institutional Review Boards: A Time for Reform* (Washington, DC: Department of Health and Human Services, Office of the Inspector General, 1998), iii.

55. NBAC, *Reports and Recommendations*, 147.

56. General Accounting Office, *Scientific Research*, 2.

57. NBAC, *Reports and Recommendations*, 143.

58. NBAC, transcript, meeting, 17 May 1997, 102, http://bioethics.georgetown.edu/nbac/ [accessed 26 Dec. 2008].

59. Senate Committee on Labor and Human Resources, *Human Research Subject Protections Act of 1997*, S. 193, 105th Cong., 1st sess., 1997.

60. House Committee on Government Reform, *Institutional Review Boards: A System in Jeopardy*, 105th Cong., 2nd sess., 1998, 50–56.

61. Ellis, interview by the author.

62. NBAC, transcript, meeting, 4 Oct. 1996, 199, http://bioethics.georgetown.edu/nbac/ [accessed 26 Dec. 2008].

63. Ellis, interview by the author.

64. "Applicability to Social and Educational Research," RG 443.

65. HHS, "Final Regulations Amending Basic HHS Policy," 8367.

66. OPRR, "Sample Multiple Project Assurance," rev. 11 Aug. 1981, IdSP.

67. Louttit, "Government Regulations," in Sieber, *NIH Readings*, 179.

68. Robert E. Windom to Charlotte Kitler, 13 Sept. 1988, RES 6-01 Human Subjects, OD-NIH.

69. Ellis, interview by the author; OPRR, "Exempt Research and Research That May Undergo Expedited Review," *OPRR Reports* 95-02, 5 May 1995, www.hhs.gov/ohrp/humansubjects/guidance/hsdc95-02.htm [accessed 22 Aug. 2007].

70. Ellis, interview by the author.

71. James Bell, John Whiton, and Sharon Connelly, *Evaluation of NIH Implementation of Section 491 of the Public Health Service Act, Mandating a Program of Protection for Research Subjects* (Arlington, VA: James Bell Associates, 1998), 28.

72. Ellis, interview by the author. In 2009, the Office for Human Research Protections took a step back, conceding that "the regulations do not require that someone other than the investigator be involved in making a determination that a research study is exempt." But it continued to recommend "that investigators not be given the authority to make an independent determination that human subjects research is exempt." OHRP, "Exempt Research Determination," 14 Oct. 2009, www.hhs.gov/ohrp/policy/ [accessed 28 Oct. 2009].

73. Ellis, interview by the author.

74. Charles A. Johnson to John R. Giardino, 4 Aug. 1999, DAR; 45 CFR 46 sec. 101(b)(4).

75. Richard E. Miller to Edward Portis, 11 Feb. 2000, DAR.

76. Ellis, interview by author.

77. Jon F. Merz, "Is Human Subjects Research (Ethics) in Crisis?" *Journal of Health Politics, Policy and Law* 25 (October 2000): 980; Eliot Marshall, "Shutdown of Research at Duke Sends a Message," *Science* 284 (21 May 1999): 1246.

78. HHS, *Protecting Human Research Subjects: Status of Recommendations* (Washington, DC: Department of Health and Human Services, Office of the Inspector General, 2000), 2; John H. Mather and Terry Hartnett, "HSP and Accreditation: Growing Pains and Successes Mark First Seven Years," *Medical Research Law & Policy Report* (20 Feb. 2008): 3; Philip J. Hilts, "New Voluntary Standards Are Proposed for Experiments on People," *New York Times*, 29 Sept. 2000.

79. Marshall, "Shutdown of Research at Duke," 1246.

80. Ellis, interview by the author.

81. Jeffrey Brainard, "Spate of Suspensions of Academic Research Spurs Questions about Federal Strategy," *Chronicle of Higher Education* (4 Feb. 2000): A29–30.

82. HHS, *Protecting Human Research Subjects*, 2.

83. President's Commission, transcript of proceedings, 12 July 1980, 329, PCSEP-GU.

84. Donna Shalala, "Protecting Research Subjects," *New England Journal of Medicine* 343 (14 Sept. 2000): 808.

85. Shalala, "Protecting Research Subjects."

86. "More Teeth for Watchdog on Research Risks," *Nature* 400 (15 July 1999): 204.

87. Brainard, "Spate of Suspensions."

88. Greg Koski, "Beyond Compliance . . . Is It Too Much to Ask?" *IRB: Ethics and Human Research* 25 (Sept.–Oct. 2003): 5.

89. NBAC, transcript of proceedings, 12 Sept. 2000, 195, http://bioethics.georgetown.edu/nbac/ [accessed 5 Aug. 2008].

90. Jessica Aungst, Amy Haas, Alexander Ommaya, and Lawrence W. Green, eds., *Exploring Challenges, Progress, and New Models for Engaging the Public in the Clinical Research Enterprise: Clinical Research Roundtable Workshop Summary* (Washington, DC: National Academies Press, 2003), 33.

91. Rick Weiss, "Research Protection Chief to Leave HHS," *Washington Post*, 17 Oct. 2002.

92. OHRP, "Federalwide Assurance for Protection for Human Subjects," version date 20 Mar. 2002, http://web.archive.org/web/20050207045125/http://www.hhs.gov/ohrp/humansubjects/assurance/filasurt.htm [accessed 10 Dec. 2008]. The prior version, dated April 2001, encouraged institutions to adopt the Belmont principles, but noted that "terms of Assurance are negotiable" (copy in the author's possession, received in response to a Freedom of Information Act request). The 2002 version dropped the promise of negotiability.

93. 45 CFR 46.103(b)(1).

94. Jonathan D. Moreno, "Goodbye to All That: The End of Moderate Protectionism in Human Subjects Research," *Hastings Center Report* 31 (May–June 2001): 10, 16.

95. Levine et al., "Political, Legal, and Moral Limits," in Knudson, *PRIM&R through the Years,* 32.

96. Carolyn Fluehr-Lobban, "Informed Consent in Anthropological Research: We Are Not Exempt," *Human Organization* 53 (Spring 1994): 4, 5, 9 n. 7.

97. Alan Sica, "Sociology as a Worldview," 254.

98. Rik Scarce, "Field Trips as Short-Term Experiential Education," *Teaching Sociology* 25 (July 1997): 219–226.

99. Mike F. Keen, "Teaching Qualitative Methods: A Face-to-Face Encounter," *Teaching Sociology* 24 (Apr. 1996): 166–176.

100. Brainard, "Spate of Suspensions."

101. Caroline H. Bledsoe, Bruce Sherin, Adam G. Galinsky, Nathalia M. Headley, Carol A. Heimer, Erik Kjeldgaard, James T. Lindgren, Jon D. Miller, Michael E. Roloff, and David H. Uttal, "Regulating Creativity: Research and Survival in the IRB Iron Cage," *Northwestern University Law Review* 101 (Special Issue, 2007): 601–602.

102. Shea, "Don't Talk to the Humans," 33.

103. Mary Jo Feldstein, "Student Fights Research Board," *Digital Missourian,* 13 May 2001, http://ludwig.missouri.edu/405/IRBstory.html [accessed 27 Oct. 2009].

104. Anonymous to Felice Levine, 19 Apr. 2000, DAR.

105. Jeffrey Brainard, "The Wrong Rules for Social Science?" *Chronicle of Higher Education* (9 Mar. 2001): A21.

106. Samuel P. Jacobs, "Stern Lessons for Terrorism Expert," *Harvard Crimson,* 23 Mar. 2007.

107. William L. Waugh Jr., "Issues in University Governance: More 'Professional' and Less Academic," *Annals of the American Academy of Political and Social Science* 585 (Jan. 2003): 91.

108. Arthur L. Caplan, "H.E.W.'s Painless Way to Review Human Research," *New York Times,* 27 Dec. 1979.

109. Mather and Hartnett, "HSP and Accreditation," 2.

110. Bledsoe et al., "Regulating Creativity," 612.

111. Jack Katz, "Toward a Natural History of Ethical Censorship," *Law & Society Review* 41 (Dec. 2007): 800.

112. Bledsoe et al., "Regulating Creativity," 616.

113. Council for Certification of IRB Professionals, *Certification Examination for IRB Professionals: Handbook for Candidates,* www.primr.org [accessed 6 Aug. 2008].

114. Collaborative Institutional Training Initiative, "About the Collaborative Institutional Training Initiative (CITI)," www.citiprogram.org [accessed 6 Aug. 2008].

115. CITI Course in the Protection of Human Research Subjects, www.citiprogram.org [accessed 30 Oct. 2006; available only to affiliates of participating organizations].

116. Council for Certification of IRB Professionals, *Certification Examination.*

117. Council for Certification of IRB Professionals, "Recertification Guidelines for Certified IRB Professionals (CIP)," www.primr.org [accessed 6 Aug. 2008].

118. Association for the Accreditation of Human Research Protection Programs, "Accreditation Principles," www.aahrpp.org [accessed 6 Aug. 2008].

119. Association for the Accreditation of Human Research Protection Programs, "Board of Directors," www.aahrpp.org [accessed 6 Aug. 2008].

120. DHEW, "Proposed Regulations," 47695.

121. Association for the Accreditation of Human Research Protection Programs, "Evaluation Instrument for Accreditation, Updated June 1, 2007," 14, www.aahrpp.org [accessed 6 Aug. 2008].

122. Susan Kornetsky, Amy Davis, and Robert J. Amdur, *Study Guide for "Institutional Review Board: Management and Function"* (Sudbury, MA: Jones and Bartlett, 2003), 37.

123. Ernest D. Prentice and Gwenn S. F. Oki, "Exempt from IRB Review," in Robert J. Amdur and Elizabeth A. Bankert, *Institutional Review Board: Management and Function* (Sudbury, MA: Jones and Bartlett, 2002), 111.

124. J. Michael Oakes, "Survey Research," in Amdur and Bankert, *Institutional Review Board*, 431.

125. 45 CFR 46.111(a)(2).

126. NBAC, transcript, 39th meeting, 6 Apr. 2000, 61–63, http://bioethics.georgetown.edu/nbac/ [accessed 24 June 2008].

127. OHRP, "OHRP Compliance Oversight Activities: Significant Findings and Concerns of Noncompliance, 10-12-2005," www.hhs.gov/ohrp/compliance/findings.pdf [accessed 22 Jan. 2009].

128. Scott Burris and Jen Welsh, "Regulatory Paradox: A Review of Enforcement Letters Issued by the Office for Human Research Protection," *Northwestern University Law Review* 101 (Special Issue, 2007): 673.

129. Ora H. Pescovitz to Kristina C. Borror, 24 July 2008, 12, www.heraldtimesonline.com/stories/2008/08/10/0808_allegations0811.pdf [accessed 10 Dec. 2008]; Nicole Brooks, "IU Research Oversight Office Has More Staff, but Projects Still Delayed," *Bloomington Herald-Times*, 8 Oct. 2008; Kevin Mackice, "Taking the H out of HCI," www.blogschmog.net [accessed 10 Dec. 2008].

130. Shea, "Don't Talk to the Humans," 28.

CHAPTER 7: THE SECOND BATTLE FOR SOCIAL SCIENCE

1. NBAC, transcript, 39th meeting, 6 Apr. 2000, 105.

2. Fluehr-Lobban, "Ethics and Professionalism in Anthropology," 63.

3. Fluehr-Lobban, "Informed Consent in Anthropological Research," 4.

4. Murray L. Wax, "Reply to Herrera," *Human Organization* 55 (Summer 1996): 238.

5. Carolyn Fluehr-Lobban, "Rejoinder to Wax and Herrera," *Human Organization* 55 (Summer 1996): 240.

6. Fluehr-Lobban, *Ethics and the Profession of Anthropology: Dialogue for Ethically Conscious Practice*, 2nd ed. (Walnut Creek, CA: AltaMira, 2002), 93.

7. John M. Kennedy, telephone interview by the author, 12 Dec. 2008.

8. American Sociological Association, *Code of Ethics and Policies and Procedures of the ASA Committee on Professional Ethics* (Washington, DC: American Sociological Association, 1999), 13–14, 20.

9. National Commission, transcript of public hearings, 653, NC-GTU.

10. Kennedy, interview by the author.

11. John Kennedy, "Hot Off the Press: Revised Draft of ASA Code of Ethics," *American Sociological Association Footnotes* (July/Aug. 1996): 6, and "ASA Members to Vote on Revised Ethics Code," *American Sociological Association Footnotes* (Mar. 1997): 7.

12. Patricia A. Adler and Peter Adler, "Do University Lawyers and the Police Define Research Values?" in Will C. van den Hoonaard, ed., *Walking the Tightrope: Ethical Issues for Qualitative Researchers* (Toronto: University of Toronto Press, 2002), 39.

13. Millstein, "DHEW Requirements," IdSP.

14. "Issues Relating to the Performance of Institutional Review Boards," NCPHS-GU; Gray, interview by author.

15. J. W. Peltason et al. to F. William Dommel, 8 Nov. 1979, box 25, Human Subjects Corresp. 1979 2/3, IdSP.

16. Richard Cándida Smith, "From the Executive Secretary," *Oral History Association Newsletter* (Winter 1989): 5.

17. Nina de Angeli Walls, "Art, Industry, and Women's Education: The Philadelphia School of Design for Women, 1848–1932," PhD diss., University of Delaware, 1995.

18. Anne Boylan to Reed Geiger, 28 Aug. 1995, DAR.

19. John Cavanaugh to Anne Boylan, 28 Aug. 1995, DAR.

20. Michael Gordon to Oral History Association discussion list, 20 Nov. 1995, DAR; Michael A. Gordon, "Historians and Review Boards," *Perspectives: Newsletter of the American Historical Association* (Sept. 1997): 35–37.

21. Margie McLellan to Oral History Association discussion list, 25 Nov. 1995, DAR.

22. Jon W. Stauff to Oral History Association discussion list, 20 Nov. 1995, DAR; Antoinette Errant to Oral History Association discussion list, 16 Apr. 1997, DAR.

23. Linda Shopes to Dale [Treleven], Don [Ritchie], and Michael [Gordon], 31 Jan. 1997, DAR.

24. American Historical Association, "Minutes of the Council Meeting, June 7–8, 1997," www.historians.org/info/annualreports/ [accessed 20 Aug. 2008].

25. Richard Cándida Smith to Stanley Katz, 19 Sept. 1997, DAR.

26. Linda Shopes and Rebecca Sharpless, letter to OPRR, 2 Mar. 1998, in the author's possession.

27. HHS, "Protection of Human Subjects: Categories of Research That May Be Reviewed by the Institutional Review Board (IRB) through an Expedited Review Procedure," *Federal Register* 63 (9 Nov. 1998): 60364–60367.

28. "OHA Involved in New Rules Affecting Academic Oral Historians," *Oral History Association Newsletter* 33 (Winter 1999): 3.

29. Shea, "Don't Talk to the Humans," 27.

30. Jonathan Knight, telephone interview by the author, 18 Sept. 2008.

31. Don Ritchie to Laurie Mercier, 3 Nov. 1999, DAR.

32. Linda Shopes to Don Ritchie, 4 June 2000, DAR.

33. Correspondent to Robert Hauck, 31 Mar. 2000, DAR. Because some scholars sent their accounts to the scholarly organizations with the expectation that their names would be kept confidential, I have omitted those names here.

34. Ed Portis to Cathy Rudder, 6 Apr. 2000, DAR; Richard E. Miller to Edward Portis, 11 Feb. 2000, DAR.

35. Correspondent to Kathleen Terry-Sharp, 12 Apr. 2000, DAR.

36. Correspondent to Felice Levine, 28 Apr. 2000, DAR.

37. ASA Web site response, 19 Apr. 2000, DAR.

38. Felice Levine to Jonathan Knight, 28 Nov. 2000, DAR.

39. My own count of fourteen sociologists' responses in the Donald Ritchie papers comes to two positive about IRBs, six negative, and the rest neutral or mixed.

40. Don Ritchie to Felice Levine et al., n.d., ca. Nov. 2000, DAR.

41. Linda Shopes to AHA Council Members, 24 May 2001, DAR.

42. American Association of University Professors, "Regulations Governing Research on Human Subjects," 363.

43. American Association of University Professors, "Institutional Review Boards and Social Science Research (2000)," www.aaup.org [accessed 17 Sept. 2008].

44. Felice J. Levine, "Comments on Proposed Revisions to the Expedited Review Categories of Research," 21 Dec. 2007, copy in the author's possession.

45. Felice J. Levine and Paula R. Skedsvold, "Where the Rubber Meets the Road: Aligning IRBs and Research Practice," *PS: Political Science & Politics* 41 (July 2008): 501.

46. Didier Fassin, "The End of Ethnography as Collateral Damage of Ethical Regulation?" *American Ethnologist* 33 (Nov. 2006): 522; Søren Holm, "The Danish Research Ethics Committee System—Overview and Critical Assessment," in NBAC, *Ethical and Policy Issues*, F-10; Robert Dingwall, "The Ethical Case against Ethical Regulation in Humanities and Social Science Research," *21st Century Society* 3 (Feb. 2008): 2; Robert Dingwall, " 'Turn Off the Oxygen . . . ,' " *Law & Society Review* 41 (Dec. 2007): 788–789.

47. Mark Israel and Iain Hay, *Research Ethics for Social Scientists: Between Ethical Conduct and Regulatory Compliance* (London: Sage, 2006), chapter 4.

48. Economic and Social Research Council, *Research Ethics Framework* (Swindon, UK: Economic and Social Research Council, 2005), 7.

49. Australian National Health and Medical Research Council, Australian Research Council, and Australian Vice-Chancellors' Committee, *National Statement on Ethical Conduct in Human Research* (Canberra: Australian Government, 2007), 7.

50. Canadian Institutes of Health Research, Natural Sciences and Engineering Research Council of Canada, and Social Sciences and Humanities Research Council of Canada, *Tri-Council Policy Statement: Ethical Conduct for Research Involving Humans, 1998 (with 2000, 2002 and 2005 amendments)* (Ottawa, ON: Interagency Secretariat on Research Ethics, 2005), i.9.

51. Economic and Social Research Council, *Research Ethics Framework*, 22.

52. Dingwall, " 'Turn Off the Oxygen,' " 789–790.

53. Kevin D. Haggerty, "Ethics Creep: Governing Social Science Research in the Name of Ethics," *Qualitative Sociology* 27 (Winter 2004): 412.

54. Carolyn Ells and Shawna Gutfreund, "Myths about Qualitative Research and the *Tri-Council Policy Statement*," *Canadian Journal of Sociology* 31 (Summer 2006): 361–362.

55. Social Sciences and Humanities Research Ethics Special Working Committee, *Giving Voice to the Spectrum* (Ottawa, ON: Interagency Advisory Panel and Secretariat on Research Ethics, 2004), 10.

56. M. H. Fitzgerald, P. A. Phillips, and E. Yule, "The Research Ethics Review Process and Ethics Review Narratives," *Ethics & Behavior* 16 (issue 4, 2006): 379.

57. Israel and Hay, *Research Ethics for Social Scientists*, 1.

58. Linda Shopes, e-mail to the author, 16 Nov. 2008.

59. NBAC, transcript, 39th meeting, 6 Apr. 2000, 142–147.

60. Linda Shopes to AHA Research Division, 24 Sept. 2000, DAR.

61. Levine to Knight, 28 Nov. 2000, DAR.

62. NBAC, *Ethical and Policy Issues*, 9.

63. Michael Carhart to Linda Shopes, 11 Dec. 2000, DAR.

64. James D. Shelton, "How to Interpret the Federal Policy for the Protection of Human Subjects or 'Common Rule' (Part A)," *IRB: Ethics and Human Research* 21 (Nov.–Dec. 1999): 6.

65. Agency for International Development, "Guide for Interpreting the Federal Policy for the Protection of Human Subjects," www.usaid.gov [accessed 12 Sept. 2008].

66. Alan Sandler to Victoria Harden, 21 Sept. 2001, DAR. When this argument was explained to Barbara Mishkin—who, as a National Commission staffer, had done much to shape the definition of research used by the regulations—she found it plausible but noted that the commission had primarily had medical research in mind when it chose the term. "When we said 'generalizable knowledge,' what we meant was you're going to do your little research project on whether or not bacterium A has any effect on this kind of population, either alone or in combination with bacterium B. And the whole purpose is to be able to contribute to medical knowledge. History—I guess it is combined knowledge. Is it generalizable? I haven't thought about that before." Mishkin, interview by the author.

67. Michael C. Carhart, "Excluding History from Bioethical Oversight," Jan. 2002, DAR; Greg Koski to Linda Shopes, Michael Carhart, Janet Golden, Jonathan Knight, and Don Ritchie, 8 Jan. 2002, DAR.

68. Shopes to Ritchie, 9 Jan. 2002, DAR.

69. Linda Shopes, comments before the National Human Research Protections Advisory Commission, 29 Jan. 2002, DAR.

70. Linda Shopes and Donald A. Ritchie to Greg Koski and Jeffrey Cohen, 4 Oct. 2002, DAR; Linda Shopes to Jonathan Knight, 4 Aug. 2003, DAR.

71. Shopes to Knight, 4 Aug. 2003, DAR.

72. Michael A. Carome, "Letter to Linda Shopes and Donald A. Ritchie, 22 September 2003," www.historians.org/press/IRBLetter.pdf [accessed 24 June 2008].

73. Linda Shopes to Don Ritchie, 19 Aug. 2003, DAR; Donald A. Ritchie and Linda Shopes, "Oral History Excluded from IRB Review," *Oral History Association Newsletter* 37 (Winter 2003): 1; Bruce Craig, "Oral History Excluded from IRB Review," *Perspectives: Newsletter of the American Historical Association*, Dec. 2003, www.historians.org/Perspectives/ [accessed 29 Oct. 2009].

74. University of California–Los Angeles, Office for the Protection of Research Subjects, "Oral History and Human Subjects Research," www.oprs.ucla.edu/human/ [accessed 8 Sept. 2008].

75. National Commission, transcript, meeting #15, Feb. 1976, 306, NCPHS-GU.

76. Michael A. Carome, "E-Mail to Lori Bross, 1 December 2003," www.nyu.edu/ucaihs/forms/oralhistory/email.php [accessed 24 June 2008].

77. Linda Shopes to Michael Carome, 19 Dec. 2003, DAR.

78. Linda Shopes to Ron Doel, 19 Jan. 2004, DAR.

79. Michael Carome to Linda Shopes, 23 Dec. 2003, DAR.

80. Michael Carome to Linda Shopes and Don Ritchie, 8 Jan. 2004, DAR.

81. Robert B. Townsend and Mériam Belli, "Oral History and IRBs: Caution Urged as Rule Interpretations Vary Widely," *Perspectives: Newsletter of the American Historical Association* (Dec. 2004): 9.

82. Robert B. Townsend, "Oral History and Review Boards: Little Gain and More Pain," *Perspectives: Newsletter of the American Historical Association* (Feb. 2006): 2.

83. OHRP, "Oral History Archive," www.hhs.gov/ohrp/belmontArchive.html/ [accessed 8 Sept. 2008].

84. Kevin Nellis, e-mail to the author, 10 Jan. 2007.

85. Columbia University, "IRB Review of Oral History Projects, 27 December 2007," www.cumc.columbia.edu/dept/irb/policies/ [accessed 8 Sept. 2008].

86. Amherst College, "IRB Policies, Procedures, and Review Guidelines," www.amherst.edu/academiclife/funding/irb/; University of Nebraska–Lincoln, "IRB Review of Oral History Projects," 1 Oct. 2008, http://research.unl.edu/orr/docs/UNLOralHistoryPolicy.pdf; University of Missouri–Kansas City, "Social Sciences IRB and Oral History," web2.umkc.edu/research/ors/Support/IRB/SS/OralHistory.html [all accessed 20 July 2009]; University of Michigan, "Activities Subject to the HRPP (July 2009)," www.hrpp.umich.edu/om/Part4.html [accessed 29 Oct. 2009]

87. Knight, telephone interview by the author.

CHAPTER 8: ACCOMMODATION OR RESISTANCE?

1. National Commission, transcript, meeting #38, Jan. 1978, 216, NCPHS-GU.

2. Stark, "Victims in Our Own Minds?" 783.

3. Dougherty and Kramer, "A Rationale for Scholarly Examination of Institutional Review Boards," 185.

4. Stuart Plattner, "Human Subjects Protection and Cultural Anthropology," *Anthropological Quarterly* 76 (Spring 2003): 287–297.

5. Joan E. Sieber, Stuart Plattner, and Philip Rubin, "How (Not) to Regulate Social and Behavioral Research," *Professional Ethics Report* 15 (Spring 2002): 1.

6. Dvora Yanow and Peregrine Schwartz-Shea, "Institutional Review Boards and Field Research," presentation at the American Political Science Association Annual Meeting, Chicago, Illinois, Aug. 30–Sept. 2, 2007, 29.

7. Charles L. Bosk, "The New Bureaucracies of Virtue, or When Form Fails to Follow Function," *PoLAR: Political and Legal Anthropology Review* 30 (Nov. 2007): 193.

8. "Communication Scholars' Narratives of IRB Experiences," 204–205.

9. Patricia A. Marshall, "Human Subjects Protections, Institutional Review Boards, and Cultural Anthropological Research," *Anthropological Quarterly* 76 (Spring 2003): 269–285.

10. NBAC, transcript, 39th meeting, 6 Apr. 2000, 104; Elisa J. Gordon, "Trials and Tribulations of Navigating IRBs: Anthropological and Biomedical Perspectives of 'Risk' in Conducting Human Subjects Research," *Anthropological Quarterly* 76 (Spring 2003): 299–320.

11. "Communication Scholars' Narratives of IRB Experiences," 220.

12. Léo Charbonneau, "Ethics Boards Harming Survey Research, Says York Professor," *University Affairs*, 6 June 2005, www.universityaffairs.ca/ethics-boards-harming-survey-research-says-york-professor.aspx [accessed 28 Oct. 2009].

13. Will C. van den Hoonaard, "Introduction: Ethical Norming and Qualitative Research," in van den Hoonaard, *Walking the Tightrope*, 11.

14. "Communication Scholars' Narratives of IRB Experiences," 204.

15. Lynne C. Manzo and Nathan Brightbill, "Toward a Participatory Ethics," in Sara Louise Kindon, Rachel Pain, and Mike Kesby, eds., *Participatory Action Research Approaches and Methods: Connecting People, Participation and Place* (Milton Park, UK: Routledge, 2007), 34; Jonathan T. Church, Linda Shopes, and Margaret A. Blanchard, "Should All Disciplines Be Subject to the Common Rule?" *Academe* 88 (May–June 2002): 62–69.

16. Linda Shopes, "Institutional Review Boards Have a Chilling Effect on Oral History," *Perspectives: Newsletter of the American Historical Association* (Sept. 2000): 6; Nancy Janovicek, "Oral History and Ethical Practice: Toward Effective Policies and Procedures," *Journal of Academic Ethics* 4 (Dec. 2006): 163.

17. "Communication Scholars' Narratives of IRB Experiences," 207.

18. Haggerty, "Ethics Creep," 403.

19. Church et al., "Should All Disciplines Be Subject to the Common Rule?"

20. Earl Lane, "AAAS Meeting Explores Ways to Improve Ethics Panels That Oversee Social Science Research," 7 Oct. 2008, www.aaas.org/news/releases/ [accessed 24 Oct. 2008].

21. Megan K. Blake, "Formality and Friendship: Research Ethics Review and Participatory Action Research," *ACME: An International E-Journal for Critical Geographies* 6 (issue 3, 2007): 417.

22. "Communication Scholars' Narratives of IRB Experiences," 207–208.

23. "Communication Scholars' Narratives of IRB Experiences," 214.

24. Sieber et al., "How (Not) to Regulate Social and Behavioral Research," 1.

25. NBAC, transcript, 39th meeting, 6 Apr. 2000, 60.

26. Gordon, "Trials and Tribulations."

27. Haggerty, "Ethics Creep," 398.

28. Adler and Adler, "Do University Lawyers and the Police Define Research Values?" 35.

29. Jack Katz, "Ethical Escape Routes for Underground Ethnographers," *American Ethnologist* 33 (Nov. 2006): 501; Church et al., "Should All Disciplines Be Subject to the Common Rule?"

30. Shea, "Don't Talk to the Humans," 29.

31. Ted Palys and John Lowman, "One Step Forward, Two Steps Back: Draft TCPS2's Assault on Academic Freedom," 15 Mar. 2009, 19, www.sfu.ca/~palys/ [accessed 9 July 2009].

32. Katz, "Toward a Natural History of Ethical Censorship," 801–803.

33. Laura Jeanine Morris Stark, "Morality in Science: How Research Is Evaluated in the Age of Human Subjects Regulation," PhD diss., Princeton University, 2006, 211.

34. Mark Kleiman, "The IRB Horror Show," The Reality-Based Community, 2 May 2009, www.samefacts.com [accessed 9 July 2009].

35. Bledsoe et al., "Regulating Creativity," 622.

36. Sue Richardson and Miriam McMullan, "Research Ethics in the UK: What Can Sociology Learn from Health?" *Sociology* 41 (Dec. 2007): 1124.

37. Adler and Adler, "Do University Lawyers and the Police Define Research Values?" 36.

38. Tara Star Johnson, "Qualitative Research in Question: A Narrative of Disciplinary Power with/in the IRB," *Qualitative Inquiry* 14 (Mar. 2008): 212–232.

39. Kathryn L. Staley, "Re: Ethics & LGBT Mental Health Service Users," H-Oralhist 15 Jan. 2007, ww.h-net.org/~oralhist/ [accessed 29 Oct. 2009].

40. Scott Atran, "Research Police—How a University IRB Thwarts Understanding of Terrorism," Institutional Review Blog, 28 May 2007, www.institutionalreviewblog.com [accessed 24 Oct. 2008].

41. Dingwall, "Ethical Case Against Ethical Regulation," 9–10.

42. Ellis, interview by the author.

43. University of California–Los Angeles, Office for the Protection of Research Subjects, "Policy Number 3: Human Subjects Research Determinations, 5 July 2007," www.oprs.ucla.edu/human/documents/pdf/3.pdf [accessed 7 Aug. 2008].

44. University of California–Los Angeles, Office for Protection of Research Subjects, "Policy Number 42: Research Involving Public Use Data Files, 17 June 2008," www.oprs.ucla.edu/human/documents/pdf/42.pdf [accessed 7 Aug. 2008].

45. Matt Bradley, "Silenced for Their Own Protection: How the IRB Marginalizes Those It Feigns to Protect," *ACME: An International E-Journal for Critical Geographies* 6 (issue 3, 2007): 340–342.

46. Bledsoe et al., "Regulating Creativity," 614.

47. Bledsoe et al., "Regulating Creativity," 620.

48. "Communication Scholars' Narratives of IRB Experiences," 222.

49. Jacobs, "Stern Lessons for Terrorism Expert."

50. "Communication Scholars' Narratives of IRB Experiences," 222.

51. Jim Vander Putten, "Wanted: Consistency in Social and Behavioral Science Institutional Review Board Practices," *Teachers College Record*, 14 September 2009, www.tcrecord.org [accessed 30 October 2009].

52. Linda C. Thornton, "The Role of IRBs in Music Education Research," in Linda K. Thompson and Mark Robin Campbell, eds., *Diverse Methodologies in the Study of Music Teaching and Learning* (Charlotte, NC: Information Age, 2008).

53. "Communication Scholars' Narratives of IRB Experiences," 206–208.

54. Sue Tolleson-Rinehart, "A Collision of Noble Goals: Protecting Human Subjects, Improving Health Care, and a Research Agenda for Political Science," *PS: Political Science & Politics* 41 (July 2008): 509.

55. Bledsoe et al., "Regulating Creativity," 619.

56. "Communication Scholars' Narratives of IRB Experiences," 224.

57. Richardson and McMullan, "Research Ethics in the UK," 1124.

58. Richardson and McMullan, "Research Ethics in the UK," 1125.

59. Mary Brydon-Miller and Davydd Greenwood, "A Re-Examination of the Relationship between Action Research and Human Subjects Review Processes," *Action Research* 4 (Mar. 2006): 123.

60. Church et al., "Should All Disciplines Be Subject to the Common Rule?"

61. Scott Jaschik, "Who's Afraid of Incestuous Gay Monkey Sex?" *Inside Higher Ed*, 14 Aug. 2007, www.insidehighered.com [accessed 28 Oct. 2009].

62. Will C. van den Hoonaard and Anita Connolly, "Anthropological Research in Light of Research-Ethics Review: Canadian Master's Theses, 1995–2004," *Journal of Empirical Research on Human Research Ethics* 1 (June 2006): 65.

63. Galliher et al., *Laud Humphreys*, 101.

64. Brydon-Miller and Greenwood, "A Re-Examination of the Relationship," 122.

65. Fitzgerald et al., "Research Ethics Review Process," 377–395; Stark, "Morality in Science." For a more detailed analysis of these writings, see Zachary M. Schrag, "Maureen Fitzgerald's Ethics Project" and "How IRBs Decide—Badly: A Comment on Laura Stark's 'Morality in Science,'" both at Institutional Review Blog, www.institutionalreviewblog.com.

66. Jenn Craythorne to Kathleen Terry-Sharp, 8 Jan. 2000, DAR.

67. "Communication Scholars' Narratives of IRB Experiences," 206.

68. "Communication Scholars' Narratives of IRB Experiences," 225.

69. Richardson and McMullan, "Research Ethics in the UK," 1122.

70. John D. Willard V, "IRB Review of Oral History," H-Oralhist, 25 Apr. 2009, www.h-net.org/~oralhist/ [accessed 28 Oct. 2009].

71. Peter Moskos, "More on IRBs," Cop in the Hood, 15 Feb. 2008, www.copinthehood.com [accessed 24 Oct. 2008].

72. Adam Hedgecoe, "Research Ethics Review and the Sociological Research Relationship," *Sociology* 42 (Oct. 2008): 880.

73. Deborah Winslow, "NSF Supports Ethnographic Research," *American Ethnologist* 33 (Nov. 2006): 521.

74. "IRBs and Behavioral and Social Science Research: Finding the Middle Ground," *AAHRPP Advance* (Winter 2008): 1.

75. Cohen, "As Ethics Panels Expand Grip."

76. Bernard A. Schwetz to Don Ritchie, 19 Apr. 2007, DAR.

77. "Communication Scholars' Narratives of IRB Experiences," 226.

78. Plattner, "Human Subjects Protection and Cultural Anthropology."

79. Charles L. Bosk and Raymond G. De Vries, "Bureaucracies of Mass Deception: Institutional Review Boards and the Ethics of Ethnographic Research," *Annals of the American Academy of Political and Social Science* 595 (Sept. 2004): 255.

80. James Weinstein, "Institutional Review Boards and the Constitution," *Northwestern University Law Review* 101 (Special Issue, 2007): 493–562.

81. Stuart Plattner, "Comment on IRB Regulation of Ethnographic Research," *American Ethnologist* 33 (Nov. 2006): 527.

82. "IRBs and Behavioral and Social Science Research," 6.

83. Plattner, "Human Subjects Protection and Cultural Anthropology," 296.

84. Plattner, "Human Subjects Protection and Cultural Anthropology," 296.

85. Levine and Skedsvold, "Where the Rubber Meets the Road," 502.

86. Kristine L. Fitch, "Difficult Interactions between IRBs and Investigators: Applications and Solutions," *Journal of Applied Communication Research* 33 (Aug. 2005): 275.

87. Charles Bosk, "The Ethnographer and the IRB: Comment on Kevin D. Haggerty, 'Ethics Creep: Governing Social Science Research in the Name of Ethics,'" *Qualitative Sociology* 27 (Dec. 2004): 417.

88. Marshall, "Human Subjects Protections," 272, 280.

89. Plattner, "Comment on IRB Regulation," 527.

90. "Communication Scholars' Narratives of IRB Experiences," 206.

91. Levine and Skedsvold, "Where the Rubber Meets the Road," 501.

92. Shelton, "How to Interpret the Federal Policy," 6.

93. Sieber et al., "How (Not) to Regulate Social and Behavioral Research," 3.

94. National Science Foundation, "Frequently Asked Questions and Vignettes: Interpreting the Common Rule for the Protection of Human Subjects for Behavioral and Social Science Research," www.nsf.gov/bfa/dias/policy/ [accessed 12 Sept. 2008].

95. "Social and Behavioral Sciences Working Group on Human Research Protections," www.aera.net/humansubjects/ [accessed 12 Sept. 2008].

96. Singer and Levine, "Protection of Human Subjects of Research," 153–154.

97. Bosk and De Vries, "Bureaucracies of Mass Deception," 250.

98. Social and Behavioral Sciences Working Group on Human Research Protections, "Risk and Harm," www.aera.net/humansubjects/ [accessed 12 Sept. 2008].

99. Social and Behavioral Sciences Working Group on Human Research Protections, "Institutional Arrangements for Reviewing Exempt, Expedited, or Other Research and Research-Related Activities," www.aera.net/humansubjects/ [accessed 12 Sept. 2008].

100. Bosk and De Vries, "Bureaucracies of Mass Deception," 251.

101. Citro et al., *Protecting Participants*, 8.

102. Citro et al., *Protecting Participants*, 35.

103. Citro et al., *Protecting Participants*, 143.

104. NBAC, *Ethical and Policy Issues*, 7.

105. Citro et al., *Protecting Participants*, 23.

106. Citro et al., *Protecting Participants*, 53.

107. Citro et al., *Protecting Participants*, 160.

108. Joan E. Sieber, "The Evolution of Best Ethical Practices in Human Research," *Journal of Empirical Research on Human Research Ethics* 1 (Mar. 2006): 1.

109. See, for example, Michael Fendrich, Adam M. Lippert, and Timothy P. Johnson, "Respondent Reactions to Sensitive Questions," *Journal of Empirical Research on Human*

Research Ethics 2 (Sept. 2007): 31–37, and Joan E. Sieber, "Protecting the Vulnerable: Who Are They?" *Journal of Empirical Research on Human Research Ethics* 3 (Mar. 2008): 1–2.

110. Philip Rubin and Joan E. Sieber, "Empirical Research in IRBs and Methodologies Usually Associated with Minimal Risk," *Journal of Empirical Research on Human Research Ethics* 1 (Dec. 2006): 3.

111. Rubin and Sieber, "Empirical Research in IRBs and Methodologies," 3.

112. Levine, "Comments on Proposed Revisions," copy in the author's possession.

113. Katz, "Ethical Escape Routes," 500.

114. Martin Tolich and Maureen H. Fitzgerald, "If Ethics Committees Were Designed for Ethnography," *Journal of Empirical Research on Human Research Ethics* 1 (June 2006): 71.

115. Linda Shopes, "Negotiating Institutional Review Boards," *Perspectives: Newsletter of the American Historical Association* (Mar. 2007): 3.

116. Richard A. Shweder, "Protecting Human Subjects and Preserving Academic Freedom: Prospects at the University of Chicago," *American Ethnologist* 33 (Nov. 2006): 516.

117. Cary Nelson, "The Brave New World of Research Surveillance," *Qualitative Inquiry* 10 (Apr. 2004): 211.

118. Edna Bonacich, "Working with the Labor Movement: A Personal Journey in Organic Public Sociology," *American Sociologist* 36 (Sept. 2005): 116; Donald L. Mosher, "Balancing the Rights of Subjects, Scientists, and Society: 10 Principles for Human Subject Committees," *Journal of Sex Research* 24 (1988): 378.

119. NBAC, transcript, 39th meeting, 6 Apr. 2000, 100; see also Dvora Yanow and Peregrine Schwartz-Shea, "Reforming Institutional Review Board Policy: Issues in Implementation and Field Research," *PS: Political Science & Politics* 41 (July 2008): 33–34.

120. Carol Rambo, "Handing IRB an Unloaded Gun," *Qualitative Inquiry* 13 (Apr. 2007): 361.

121. Jonathan Moss, "If Institutional Review Boards Were Declared Unconstitutional, They Would Have to Be Reinvented," *Northwestern University Law Review* 101 (Special Issue, 2007): 804.

122. Bledsoe et al., "Regulating Creativity," 606.

123. Nelson, "Brave New World of Research Surveillance," 216.

124. Bosk, "New Bureaucracies of Virtue," 198.

125. Scott Burris, "Regulatory Innovation in the Governance of Human Subjects Research: A Cautionary Tale and Some Modest Proposals," *Regulation & Governance* 2 (Mar. 2008): 80.

126. "Communication Scholars' Narratives of IRB Experiences," 227.

127. Will C. van den Hoonaard, "Is Research-Ethics Review a Moral Panic?" *Canadian Review of Sociology and Anthropology* 38 (Feb. 2001): 19–36; M. H. Fitzgerald, "Punctuated Equilibrium, Moral Panics and the Ethics Review Process," *Journal of Academic Ethics* 2 (Dec. 2005): 315–338.

128. Debra Viadero, "Security Checks of U.S. Education Contractors to Change," *Education Week* (2 Apr. 2008): 8.

129. Alice Dreger, "The Vulnerable Researcher and the IRB," *Bioethics Forum*, www.thehastingscenter.org/BioethicsForum/ [accessed 6 Oct. 2008].

130. Bosk, "New Bureaucracies of Virtue," 194.

131. HHS, *Protecting Human Research Subjects*, 3.

132. NBAC, *Ethical and Policy Issues*, 7, 9.

133. "Communication Scholars' Narratives of IRB Experiences," 221; Daniel Bradburd, "Fuzzy Boundaries and Hard Rules: Unfunded Research and the IRB," *American Ethnologist* 33 (Nov. 2006): 492–498.

134. Shea, "Don't Talk to the Humans," 32.

135. Bonacich, "Working with the Labor Movement," 116.

136. Shea, "Don't Talk to the Humans," 32.

137. Haggerty, "Ethics Creep," 410.

138. "Michaela," comment on "Embodied Ethics Oversight," 5 May 2009, Culture Matters, http://culturematters.wordpress.com (4 June 2009).

139. Bledsoe et al., "Regulating Creativity," 625.

140. Shea, "Don't Talk to the Humans," 30–31.

141. "Communication Scholars' Narratives of IRB Experiences," 210.

142. "Communication Scholars' Narratives of IRB Experiences," 214.

143. Mark H. Ashcraft and Jeremy A. Krause, "Social and Behavioral Researchers' Experiences with Their IRBs," *Ethics & Behavior* 17 (Jan. 2007): table 3; Viadero, "Security Checks"; Richardson and McMullan, "Research Ethics in the UK," 1122.

144. Shweder, "Protecting Human Subjects," 509.

145. John A. Robertson, "The Social Scientist's Right to Research and the IRB System," in Beauchamp et al., *Ethical Issues in Social Science Research*, 356.

146. Philip Hamburger, "The New Censorship: Institutional Review Boards," *Supreme Court Review* (Oct. 2004): 281. See also Kerr, "Unconstitutional Review Board?" 393–447.

147. Gunsalus et al., "Illinois White Paper."

148. American Association of University Professors, "Research on Human Subjects: Academic Freedom and the Institutional Review Board," www.aaup.org/AAUP/comm/rep/A/humansubs.htm [accessed 28 Oct. 2009].

149. Katz, "Ethical Escape Routes," 504.

150. NBAC, transcript, 39th meeting, 6 Apr. 2000, 76–77.

151. Levine and Skedsvold, "Where the Rubber Meets the Road," 504.

152. Levine and Skedsvold, "Where the Rubber Meets the Road," 503.

153. Greg Downey, "Dr. Zachary Schrag on Ethics, IRB & Ethnography," Culture Matters, 20 Aug. 2007, http://culturematters.wordpress.com [accessed 2 Nov. 2008].

154. "Policy Regarding Human Subject Research in the Sociobehavioral Sciences," *University of Pennsylvania Almanac*, 3 Oct. 2006, www.upenn.edu/almanac/ [accessed 2 Nov. 2008].

155. Tuskegee Syphilis Study Ad Hoc Advisory Panel, *Final Report of the Tuskegee Syphilis Study Ad Hoc Advisory Panel* (Washington, DC: Department of Health, Education, and Welfare, 1973), 24.

156. National Commission, transcript, meeting #24, Nov. 1976, 175, NCPHS-GU.

CONCLUSION

1. Venkatesh, *Gang Leader for a Day*, 206.

2. Ellis, interview by the author.

3. Gray, interview by the author.

4. President's Commission, transcript of proceedings, 12 July 1980, 333–340, PCSEP-GU.

5. Jeffrey Cohen, "PRIM&R Thoughts," 21 November 2008, HRPP Blog, http://hrpp.blogspot.com [accessed 24 Dec. 2008].

6. Robert J. Levine, "Empirical Research to Evaluate Ethics Committees' Burdensome and Perhaps Unproductive Policies and Practices: A Proposal," *Journal of Empirical Research on Human Research Ethics* 1 (September 2006), 3.

7. Secretary's Advisory Committee on Human Research Protections, transcript, 16th meeting, 16 July 2008, 263–264.

INDEX